T0320758

Depression in Schizophrenics

Depression in Schizophrenics

Edited by

Richard Williams and
J. Thomas Dalby

Calgary General Hospital
and University of Calgary
Calgary, Alberta, Canada

Plenum Press • New York and London

Library of Congress Cataloging in Publication Data

World Psychiatric Association Symposium on Depression in Schizophrenics (1988: Calgary, Alta.)
   Depression in schizophrenics / edited by Richard Williams and J. Thomas Dalby.
     p.   cm.
   "Proceedings of a World Psychiatric Association Symposium on Depression in Schizophrenics,
held July 6–7, 1988, in Calgary, Alberta, Canada"—T.p. verso.
   Includes bibliographical references and index.
   ISBN 0-306-43240-4
   1. Schizophrenia—Congresses. 2. Depression, Mental—Congresses. I. Williams, Richard,
1950–   II. Dalby, J. Thomas. III. Title.
   [DNLM: 1. Depressive Disorder—congresses. 2. Schizophrenia—congresses. QM 203 W92763d
1988]
RC514.W68   1988
616.89′82—dc20
DNLC/DLC   89-16146
for Library of Congress   CIP

Proceedings of a World Psychiatric Association Symposium on
Depression in Schizophrenics, held July 6–7, 1988,
in Calgary, Alberta, Canada

© 1989 Plenum Press, New York
A Division of Plenum Publishing Corporation
233 Spring Street, New York, N.Y. 10013

Printed in the United States of America

PREFACE

The chapters of this volume were originally presented at a symposium on "Depression in Schizophrenics" held at the University of Calgary, Health Sciences Centre on July 06 and 07, 1988. It was the intent of the organizers to draw together leading international researchers to address, in a comprehensive fashion, the persisting problem of depression in schizophrenic individuals.

As many of the authors point out, depression was clearly identified as a central problem in schizophrenia by the pioneers of psychiatry. Their wisdom and clinical acumen was lost for a time to be only recently re-discovered. Their insights must now be integrated with modern taxonomic systems, evolving etiological models and methods of assessment. With increased recognition of the problem of depression in this population we must also examine appropriate methods of ameliorating the suffering of those afflicted.

The conference was divided into sections (theoretical aspects, phenomonology, suicide and treatment) and the chairmen of these sections have given brief introductions. Necessary overlap of topics and methods provided a convergent and validating examination of empirical results which may serve to direct continuing studies. As well we hope that thoughtful review may directly influence clinical practice and thereby improve the quality of life for schizophrenic patients.

R. Williams
J.T Dalby

ACKNOWLEDGEMENTS

The symposium on which this book is based was co-sponsored by a number of organizations including the Department of Psychiatry, University of Calgary; the World Psychiatric Association; the World Psychiatric Section on Psychiatric Rehabilitation; the Alberta Psychiatric Association and the Alberta Friends of Schizophrenics. A special note of thanks is extended to Merrell Dow Pharmaceuticals (Canada) Inc. who provided a grant in aid of continuing medical education.

The impetus and organization for the conference was directed by the Schizophrenia Research Interest Group at the University of Calgary including: Don and Jean Addington, Mary Anne Beyfus, Tom Dalby, Gerry McDougall, David Romney, Di Vosburgh and Richard Williams

Ms. Jocelyn Lockyer of the Office of Continuing Medical Education at the University of Calgary provided invaluable assistance in the arrangements for the symposium. Ms. Caroline Collins of the same office contributed her patient aid by typing and proofreading the entire text.

Finally we offer our sincere appreciation to each of the programme participants.

CONTENTS

Section 1

THEORETICAL ASPECTS

INTRODUCTION

Richard Williams

Man has always classified his external and internal worlds in an attempt to understand each to a greater degree. In the 1800s classification systems for animal and plant species were refined to an extreme degree such that the system has persisted even when new evidence suggested change in subgrouping for some animals. Classifying the internal world of mental processes has proved less easy; the early organic/functional dichotomy was found to be unhelpful, for as mental disorders become better understood they tended to shift from the functional to the organic group. For example, epilepsy, G.P.I. and now even schizophrenia and affective disorders are being shown to be associated with organic changes. Whether these biological changes are the cause of, or result from, the disorder, has still to be elucidated.

Dr. Brockington's chapter is a provocative challenge to psychiatric investigators to move outside of the restricting and constricting confines of ICD-9 and DSM-III-R categories to look at major psychiatric disorders along dimensions. Walton and Presly (1) have shown that personality types do not fit neatly into categories but can be understood on a multidimensional model; perhaps it is now time for other mental disorders to be viewed in a similar way. Dr. Brockington discusses two phenomenological dimensions (one being delusions or autism; the second verbal hallucinosis and passivity) and a third behavioral dimension (negative symptoms or defect state) and it is Dr. Flor-Henry who introduces the fourth dimension of organic changes. It is clear that central laterality is an important element in understanding the pathophysiology of individual patients, as it can be related to treatment responsiveness (demonstrated in his case example). The introduction of Nuclear Magnetic Resonance Imaging (NMRI) techniques has indicated that apparently different types of abnormality (hypo- and hyper-frontality) (2) within patients diagnosed according to strict research criteria as having schizophrenia; whilst Dr. Flor-Henry attempts to explain all this within a unitary theory, one must consider the possibility of different contributing factors producing different organic changes although sharing a final common phenomenological pathway.

The genetic component to mental illness is clearly reviewed by Dr. McGuffin and colleagues. Tim Crow (3) has long postulated the inner-relatedness of affective illness and schizophrenia, and whilst not originating the "continuum of psychosis" hypothesis, has eloquently developed the theory by suggesting that _genes_ predispose to psychosis which may breed true in a family or be altered in clinical presentation by changes in the virogene due to replications within the genome. This, he argues, could also account for a "season of birth" effect. Bassett et al (4) identified a family in which schizophrenia was associated with facial dysmorphism and other physical anomalies; and in which there was abnormality of chromosome 5 in those family members affected but not in the unaffected members. Sherrington et al., (5) have reported on the localization of a susceptibility locus for schizophrenia on chromosome 5. However, in the same issue of _Nature_, where this report appeared, Kennedy et al (6) in a Swedish pedigree study failed to find evidence for chromosome 5 linkage. It is clear that techniques for restricted fragment length polymorphisms (RFLPs) will add to our knowledge of the psychoses but it would be a foolish researcher who looked for one defect to explain all cases of schizophrenia. As high technology moves researchers away from phenomenological considerations of depression and schizophrenia we must not at our own peril (as Dr. Brockington suggests) fail to integrate the dimensional aspects of each patient, and equal weight must be given to organic findings, behavioral changes and phenomenology; rather than to currently define research criteria as a priori descriptions of either schizophrenia or depression.

REFERENCES

1. Walton HJ, Presly AS: Use of a category system for diagnosis of abnormal personality. Br J Psychiatry 122: 259-268, 1973.
2. Flor-Henry P: Cerebral dynamics, laterality and psychopathology. In Cerebral Dynamics, Laterality and Psychopathology. Edited by R. Takahashi, P. Flor-Henry, J. Grugelier, S-I Niwa, Amsterdam, Elsevier Science Publishers.
3. Crow TJ: The continuum of psychosis and its implications for the structure of the gene. Br J Psychiatry 149: 419-429, 1986.
4. Bassett AS et al: partial trisomy chromosome 5 cosegregating with schizophrenia. Lancet 1: 799-801, 1988.
5. Sherrington R et al: Localization of a susceptibility locus for schizophrenia on chromosome 5. Nature 366: 164-167, 1988.
6. Kennedy JL et al: Evidence against linkage of schizophrenia to markers on chromosome 5 in a Northern Swedish pedigree. Nature 366: 167-170, 1988.

A CRITICAL VIEW OF THE CONCEPT OF SCHIZOPHRENIA

Ian F. Brockington

University of Birmingham
Birmingham, England

From my first encounter with psychiatry, I have always regarded 'schizophrenia' as an improbable entity. Yet this idea has become more and more dominant in our profession, powerful in the clinic, the law courts and the research councils.

My own empirical work in the field of nosology was carried out as part of an investigation of "schizoaffective disorders". Between 1972 and 1975 (over 12 years ago) I worked on this subject with Professor Robert Kendell, and in the next 7 years we analysed and published our results. We collected 108 patients (76 of whom had the picture of schizoaffective depression, and 32 schizoaffective mania) by screening 3,800 admissions to the Maudsley, Bethlem and St. Francis Hospitals, they were entered into drug trials and followed up 1-4 years later. We also collected a series of 119 psychotic first admissions and followed up the 134 patients in the Netherne series of consecutive psychotic admissions. The conclusion of all this was that schizomania is roughly equivalent to mania, which was a minor but helpful revision of nosological thinking at that time; as for schizoaffective depression, it was so heterogeneous in demography, symtomatology, course and treatment response, that it amounted to an unselected collection of psychotic patients. The concept of schizoaffective disorders makes no great contribution to classification or clinical practice.

A by-product of the study was a collaboration with Professor Carlo Perris in Umea, clarifying the notion of cycloid psychosis, a much more useful concept than schizoaffective disorder. This study of the schizoaffective and cycloid psychoses was for me an apprenticeship in the methodology of clinical research, and left me with a dissatisfaction with the concept of schizophrenia, which is the root of our difficulty in classifying the psychoses.

Schizophrenia's history is one of semantic wrangles. The best efforts of experts has not achieved a consensus on its meaning. The concept has been geographically and historically unstable. Its interpretation has been strongly influenced by local tradition, and is subject to social rather than scientific constraints. Uniformity of definition has been imposed by professional organizations (such as the American Psychiatric Association) within the bounds of their authority but this is limited to their own parish.

When one looks for the reason for the failure to clarify this concept, there are two:

First, Kraepelin's hypothesis of a unitary disease entity has not been confirmed. Not a single pathognomonic symptom or marker has been discovered, and discriminant function analysis has failed to find a natural boundary between schizophrenia and the affective psychoses. The clinical spectrum of the psychoses is a continuum with an abundance of intermediate forms.

Secondly, schizophrenia lacks a central principle which all recognize as the essence of the idea; it has not one but three defining principles:

1. Kraepelin recognized convergence to a defect state "with a peculiar destruction of the internal connections of the psychic personality with the most marked damage to the emotional life and volition" - the familiar idea of a final common path.

2. Others have emphasized the schizm or fission within the psyche ("a fractionation of the mind, a true dualism in the same individual"), which is at the heart of the Schneiderian concept of schizophrenia.

3. Neither of these exhaust the phenomena of schizophrenia, because there is also autism, i.e., the aloneness, the excessive individualism and idiosyncrasy, absorption with a private world without any attempt to reconcile it with consensual reality.

These principles are all completely different, and thus the richness of the psychopathology of schizophrenia, which Kraepelin and Bleuler described so vividly and which gives the 'disease' its fascination and appeal, is its greatest weakness, denying any possibility of bringing the concept to a clear focus.

The idea of schizophrenia has permitted advances. Notably the study of these patients has disclosed the hereditary transmission (different from that of affective disorders) of lifelong social disabilities which include much of schizophrenia in addition to a spectrum of personality disorders. But patients with this so-called 'disease entity' are so diverse both in clinical picture and course that researchers have been served with hopelessly heterogeneous groups of patients. Meanwhile the grandeur of the big idea has diverted attention from less common, more homogeneous entities whose study is more promising. Worst, it has stifled clinical curiosity. Psychiatry as a clinical science holds privileged access to patients. The task of observing clinical phenomena urged on us by Sir Thomas Lewis is still in its early stages, yet with notable exceptions there has been a widespread loss of inquisitiveness about mental phenomena. Schizophrenia has given the illusion of understanding, an unjustified complacency, a spurious authority.

In drawing attention to 'dementia praecox' as a final common pathway, and making it the defining principle, Kraepelin led psychiatry down the wrong path. Consequences are important, more so in psychiatry than in somatic medicine, because our patients have to live out a normal lifespan with them, but the main thrust must be towards the pathogenesis and origins of psychopathology, to search out the relationship between challenging events and cerebral lesions and the normal structure (chemistry, physiology, psychology) of the mind and brain. The 'disease entity' of schizophrenia is too crude an oversimplification to help in this task.

Kraepelin would have seen that 100 years is long enough for an idea introduced "for the purpose of preliminary enquiry", and, being the innovator he was, would have developed a more complex and diversified classification. His idea of schizophrenia as a disease entity may have been misunderstood. The model of a disease with a single cause, of which the best examples are found in the infectious diseases, nutritional disorders and inborn errors of metabolism seems inappropriate for one of the most prevalent disorders in psychiatry. The alternative model of organ failure is worth considering. When the kidney fails, several different dysfunctional processes are released, including waste product retention, anaemia, hypertension and protein loss. When the lungs are diseased there is anoxaemia, carboaemia and pulmonary hypertension overloading the right heart. When the heart fails there is low cardiac output, high venous pressure and oedema. Thus the failure of a bodily structure, from whatever cause (and there are many for each organ) sets in train a variety of harmful chain reactions. Kraepelin must have had this in mind when he defined dementia praecox in terms of its final common path. In the realm of the psychoses, the structure concerned is not physical but psychological and is not at present well-defined. It is a structure of ideas, concerned with personal identity, and the nature of the social world, with the stream of thought and the direction of its contents, with emotional expression, self-control and communication. Failure of psychosocial development sets in motion a number of distinct forms of dysfunction, activated to differing extents in different individuals, and this results in the clinical spectrum we see in our patients. The task of psychiatry is to identify, isolate, quantify and analyse these vectors independently.

The formation of delusions is one axis, corresponding to the definitional theme of 'autism'. Delusions may be the only symptom of a mental illness, as in the paranoid or simple delusional psychoses. The belief that they depend partly on personality is enshrined in Kretchmer's work on the sensitive psychoses, and the role of stress is asserted by the influential Scandinavian school of thinking on the reactive or psychogenic psychoses. The work of the

Americans, McCabe and Kendler has supported the claim that these disorders are a 'third psychosis'. Delusions are complex. There are many variations in content, some of which have their own predisposing factors and precipitants, as well as different effects on perception and conduct. There is no clear division between delusional thinking and normal thinking. Delusions resemble religious and political belief systems in their support to self-esteem. They vary in intensity along a number of parameters, including preoccupation, affective response, conviction, systematization, extension and distance from shared beliefs. French psychiatric thinkers of the early years of this century (Serieux and Capgras, Ballet, Dupres and Logre, De Clerambault) argued that there are different mechanisms at work in the formation of delusions. First, there is the gradual evolution of a system of ideas through a process of reasoning, via the interpretation and misinterpretation of events; here a number of factors may contribute to the evolution of the system, including the loss of corrective feedback in those isolated by deafness or linguistic and cultural isolation, and the need to resolve debilitating conflict and ambiguity. Secondly, delusions form to explain overwhelming and perplexing experiences, as in Ballet's hallucinatory psychosis. Thirdly, they form under the force of overwhelming emotions, which can neither be contained nor discharged. Finally, they form as corruption of imagination, in which all attempts to distinguish between wish-fulfilling fantasy and reality are abandoned (la folie d'imagination). To this list one can add the misinterpretation of the world by a cerebrum fragmented by widespread cerebral dysfunction (delirium). The effect of pharmaceutical agents like amphetamines, and the influence of major mood disorders such as melancholia and mania also must be considered. Once initiated, delusions are maintained by the inertia of all ideas, by the need for self-consistency and by a number of vicious circles including social withdrawal and the loss of social effectiveness, hypervigilance, overarousal and behaviour provoking confirmatory responses from others. Although the loss of the critical faculty and failure to keep in line with consensual reality is common to all, there are many types and causes of delusions, and it follows that they are themselves non-specific and cannot identify a single disease process.

The second major symptom group in schizophrenia is that of verbal hallucinosis and passivity. This is the dimension emphasized by De Clerambault (in his 'automatisms') and Schneider (in his first rank symptoms), and used by Wing and the authors of the Research Diagnostic Criteria and 3rd Edition of the American Psychiatric Association Diagnostic and Statistical Manual as the main source of their 'characteristic' or 'nuclear' symptoms. It is first necessary to clarify (following Seglas) the distinction between verbal and non-verbal hallucinosis. Non-verbal hallucinosis is closely entwined with preoccupying ideas, which play an important part in the translation of sensory experience into perception by determining the focus of attention through expectant scanning of the environment. Verbal hallucinosis, on the other hand, is not part of the dominant preoccupation, but intrudes upon it. As Ey has explained, it is to be understood in the context of man "living in a world of words". It is a form of thought intrusion, to be compared with those which follow stressful events (researched by Horowitz), and obsessional phenomena. Verbal hallucinosis is not synonymous with auditory hallucinosis, since some auditory hallucinations are not verbal (e.g. musical hallucinations), and some verbal hallucinations, are not auditory (e.g. Seglas' psychomotor hallucinations, and those experienced in the tactile mode by the blind, or in the visual mode by the deaf). Auditory hallucinations have something in common with echo phenomena, and the experience of thought, emotion or action as the work of an alien focus within or out with the person himself. The common element is schizm or fission within the personality, in which mental events are experienced as out of the reach of the will, and without the subjective sense of belonging to the self. There is also, in auditory hallucinosis, the addition of qualities of externalization and vocalization to thought which is hard to relate to normal thinking, though it occurs in altered states of consciousness such as the hypnogogic state. I wish to stress the difference between these phenomena and delusions, even though the latter may form to explain these experiences, or under the influence of their contents. Delusions and verbal hallucinations are often found together in our patients, but this should not deflect us from the aim of understanding their separate pathogenesis, any more than the frequent association of depression and anxiety should persuade us that these are the same emotion.

The third parameter of schizophrenia is concerned with 'negative symptoms' or the 'defect state', which Kraepelin used as his defining principle for dementia praecox. In more recent times a number of people have stressed the complexity of this group of phenomena. They include social and intellectual deficits which were part of the subjects' premorbid functioning, and others which are secondary to active symptoms or treatment. I will discuss them under the separate headings of social handicap, apathy, affective blunting and thought disorder.

Social handicap may be due to shyness, envy and spite, low self-esteem, feelings of unpopularity, lack of social graces and skills, or social sensitivity (including minor paranoid states such as dysmorphobia), or they may be a form fruste of early infantile autism (Asperger's syndrome); other individuals are 'schizoid' in the sense of lacking interest in social relationships, with excessive abstract preoccupations, yet others are 'schizotypal' in the sense of having minor idiosyncrasies of thought and language. Defects of performance in school and employment may be due to minor intellectual handicaps or to social sensitivity. After the onset of the illness, and particularly the interlude of hospitalization, other complications set in: relationships and mutual respect may be shattered. There may be depression (common in all psychoses), and a loss of morale, status, and the independence which we labour to achieve during a long childhood.

In addition to these factors which are not specific to any psychosis, there has been claimed to be a more specific deficit of volition. In so far as this is not due to depression and the atrophy of confidence, or loss of roles and a sense of purpose, or the sedative effects of phenothiazines, one would look again at the influence of delusions. Apathy may be associated with grandiose delusions, in terms of which all energetic action is without purpose. As for blunting of affect, those with chronic illnesses tend to stare in a watchful and disconcerting way which has been interpreted as a lack of emotional response. Their emotions are frequently normal in the sense of showing normal anger, fear, depression and amusement but sometimes appear out of keeping with the consensual view of the circumstances and are labeled 'inappropriate'; or patients may show a lack of concern where concern is warranted. However, these responses may be appropriate to the patient's construction of the situation i.e., secondary to delusions.

What about thought disorder? This is again a complex group of disorders, which has been analysed by Andreason. Before diagnosing any form of thought disorder, one must exclude difficulties of self-expression due to low intelligence (with redundancy and poor grammar), and the use of a foreign language or patois. Then there are forms of thought disorder often seen in manic patients in which there is a ready deviation from the topic along lines of free association, due to distractibility; when extreme, manic speech may be so disorganized as to resemble that of the most severe schizophrenic patients. Patients under high emotional drive (e.g., due to fear or fury) may be incoherent because the pressure to express themselves outstrips their power to impose language rules, or because several lines of thought compete and become interfused. None of these forms of incoherence have any specificity for schizophrenia. Andreason has emphasized 'alogia' or 'poverty of content of speech' as a negative symptom, but from her examples I am not convinced that this differs from unskilled redundant self-expression. 'Verbigeration' is a term given to endless repetition of socially meaningless phrases seen in catatonia and beautifully described by Kahlbaum; it seems to be the verbal equivalent of stereotypes. 'Incoherent speech', with 'knight's move' is like manic speech, except that the train of thought cannot be followed by the observer. Finally there is the use of stock words. This is not the paralogia of Wernicke's dysphasia in which a high proportion of words used are meaningless without any repetitive pattern, but the repetition of the same words again and again (either words using general use of neologisms), perhaps because the words in question carry a portmanteau of meanings. These last two phenomena of defective communication, seem to be manifestations of the highly idiosyncratic thought patterns of an individual living autistically in his own system of private ideas, who expresses his preoccupations in his own private language, without striving to convey his meaning. Thought disorder is complex, and the more specific elements come back to the idiosyncrasy which underlies the formation of delusions, and of which they are the extreme manifestation. Thus, if we exclude hebephrenia (which may be a separate entity), the core features of the defect state can be interpreted as derivative of the theme of autistic or delusional thinking.

In presenting this more complex model of the chronic psychoses, in which the clinical spectrum results from the interaction of a skein or distinct morbid processes, I want to distance myself from the tradition of the 'unitary psychosis'. I do not believe that there is one psychosis (i.e., one morbid process at work), nor do I see the usefulness of emphasizing the clinical continuum. The task of psychiatry is to analyse this continuum, and tease out the factors at work - identify, define and measure them, and determine their causes and effects.

The model presented here, in which the chronic psychoses are seen as the sum total of an array of distinct but interrelated morbid processes, will be less easy to investigate and refute than the dichotomies advanced by Crow (type 1 and type 2 schizophrenia) and Murray (familial and sporadic - organic schizophrenia). These models have the merit of simplicity, and will appeal to those who can believe that such diverse clinical phenomena are the symptoms of a unitary disease. But psychiatry (if it is to command respect in the medical profession and intelligentsia) must use a classificatory model which is credible, and these formulations in terms of one or two entities are not believable except as heuristic oversimplifications.

The testing of such a complex model is a formidable challenge. The following research strategies are suggested:

1. Instead of reducing clinical complexities to diagnostic categories, they should be analysed into the component factors, each of which is quantified. The factors are then entered into multiple regression equations as independent variables.

2. Instead of concentrating on the study of the most severe and florid cases (those with the full spread of 'schizophrenic' symptoms), we search for and study patients who manifest only one component of symptomatology. This makes the study of patients with hallucinosis or echo phenomena, or monosymptomatic delusions, or social handicaps without psychotic symptoms, of particular value.

3. We pay more attention to the detailed study of individual patients, using every form of investigation to bring into the clearest focus the interaction between factors in the genesis and evolution of the psychosis.

In summary I believe it is important to loosen the grip which this notion of schizophrenia has on all our minds. Science is a structure of ideas representing the external world. It is clear and powerful when each idea accurately mirrors a single aspect. Its enemy is an imprecise terminology. Schizophrenia cannot be brought into sharp focus because it includes several different essential ideas. It does not help us to understand what is wrong with our patients, but is merely a cloak for ignorance. It is an obstacle to progress in the laboratory because it presents scientists with a portmanteau of variform patients. It stifles the curiosity which is the spring of new knowledge. Three generations ago, psychiatry was illuminated by the two entities principle, and galvanized into fresh initiatives, but it is now hobbled by the same ideas. In a clinical discipline, nosology is paramount, and psychiatry needs to be released from the straight jacket of this deeply entrenched idea into a new vigour and freedom of clinical enquiry.

GENETICS AND AFFECTIVE CHANGES IN SCHIZOPHRENIA

Peter McGuffin, Anne E. Farmer, Ian Harvey, and
Maureen Williams

University of Wales College of Medicine, Cardiff,
Wales, and London England
King's College, University of London

One of the pioneers of psychiatric genetics, Luxenburger (quoted in Jaspers)(1) remarked, that for the purpose of research 'schizophrenia' should be regarded as no more than a useful working hypothesis. It remains essential to retain this principle especially in the current era of research when operational diagnostic criteria such as DSM-III (2) with all their proven reliability and appearance of certainty can easily mislead us into the belief that we really know what schizophrenia is. The enhancement of reliability has of course been a boon to psychiatric research, and it is scarcely surprising that the use of explicit criteria has now become virtually mandatory in respectable scientific journals when schizophrenia is the topic. But, in addition to a false sense of security and certainty, operational definitions have two other drawbacks. First, many alternative sets of criteria are available, most with proven reliability, but when a variety of definitions is employed to classify the same sample of patients there is frequently poor concordance over which subjects are or are not diagnosed suffering from schizophrenia. As Brockington et al (3) have put it, the previous state of inarticulate confusion in the diagnosis of schizophrenia has been replaced by a 'babble of precise but differing formulations of the same concept'. The second problem is that some workers have come to question virtually any finding established before the comparatively recent introduction of operational definitions. Therefore before focussing on affective change it is worth giving some brief consideration to the genetics of schizophrenia in the context of modern diagnostic criteria.

Modern Diagnostic Criteria and Genetics

Traditionally, one of the most constant aetiological clues concerning schizophrenia has been its tendency to run in families and this has been amply confirmed in many studies (4). Surprisingly, two groups of workers (5,6) produced results which appeared to challenge this view by failing to find that schizophrenia was common among the first degree relatives of schizophrenics. Both groups of authors concluded that all previous research was probably mistaken and that there was no evidence that schizophrenia was genetic once a 'proper' definition had been applied. Such conclusions of course overlook the fact that operational definitions, although reliable, are of uncertain validity and it seems likely that a combination of methods of low sensitivity, inadequate sample size and restrictive diagnostic criteria conspired to produce falsely 'negative' results. Fortunately, other workers have completed studies with larger samples, more satisfactory methodologies and producing positive results. For example, personal interviews and assessments of clinical records of 375 first degree relatives of schizophrenics and 543 relatives of controls were carried out by Tsuang et al (7). They found, using modified St. Louis criteria (8) that the morbid risk of schizophrenia in first degree relatives of index cases was 5.5% compared with .6% in the relatives of controls. Subsequent reassessment using DSM-III criteria produced essentially similar results. Using Feighner-like criteria, Guze et al (9) carried out a follow-up and family study in St. Louis with 'blind' diagnoses based on personal interviews with relatives. A significant excess of schizophrenia was found in the relatives of schizophrenics compared with the relatives of other psychiatric patients. Baron et al (10) have also added to the evidence that schizophrenia does not cease to be familial once it has been narrowly and explicitly defined.

Demonstration of familiality does not, of course, confirm heritability. Indeed schizophrenia whether defined using a broad clinical method or restrictive research criteria rarely follows regular Mendelian segregation patterns. The evidence that it is a genetic syndrome derives from the natural experiments afforded by twin birth and adoption. No new systematic twin studies have been carried out and published since the advent of operational diagnostic criteria. However, the material collected by Gottesman and Shields (11) based on the Maudsley Twin Register, has provided sufficient detail for us to perform a recycling exercise and to reclassify probands and cotwins blind to their identity, zygosity and previously published diagnoses. Some of the important results are shown in Table 1 where it is clear that modern North American criteria for schizophrenia define a condition which shows a substantially greater concordance in monozygotic (mz) than in dizygotic (dz) twins producing an estimated heritability of around 80%. By contrast, using Schneiderian first rank symptoms alone to define the disorder proved much less satisfactory with a greatly reduced number of probands qualifying for the diagnosis and yielding an effectively zero heritability.

An adoption study is currently in progress in Finland employing modern research criteria but so far only preliminary results are available. Kendler and colleagues (17) have also performed a recycling experiment and blindly diagnosed the greater Copenhagen sample of Kety's (18) adoptees family study. Among the 69 biological relatives of adoptees who had diagnoses within the schizophrenia spectrum, 15 (22%) had similar diagnoses compared with only 3 of 137

Table 1   Twin concordance and heritabilities of operational criteria for schizophrenia [data from McGuffin et al (12); Farmer et al (13)]

Criteria	MZ twins			DZ twins			
	N pro-bands	Concor-dance %	r	N pro-bands	Concor-dance %	r	$h^2\pm$(SE)
Spitzer et al (14)							
Broad	22	45.5	0.86	23	8.7	0.45	0.83±.4
Narrow	19	52.6	0.90	21	9.5	0.48	0.83+.4
Feighner et al (8)							
Probable	21	47.6	0.88	22	9.1	0.46	0.84+.4
Definite	19	47.4	0.88	18	11.1	0.52	0.72+.4
DSM-III (2)	21	47.6	0.87	21	9.5	0.44	0.85+.4
Schneider (15)	9	22.2	0.74	4	50.0	0.91	0 +.49

Concordance is expressed probandwise

r = correlation in liability

h^2= approximate broad heritability

(2%) control relatives. Taken together therefore, the evidence from family, twin and adoption studies confirms that the working hypothesis of 'schizophrenia' reformulated operationally has an important genetic basis.

Flies in the Kraepelinian Ointment

For most of the brief history of psychiatric genetics, schizophrenia and manic depressive illness have generally been considered under separate headings. The functional psychoses have been covered by two working hypotheses and these have been kept more or less distinct. However, as will be discussed throughout this volume, it is often difficult to accept this status quo complacently. There are two conspicuous varieties of fly in the ointment. The first is the patient who appears to have one form of the two major Kraepelinian psychoses (19) while one or more close relatives has the other. The second is the patient who presents with both schizophrenic and prominent affective features. The overlap in familial distribution has long been recognized. For example, Slater (20) found that depression was more common in the parents of his schizophrenic twins than was schizophrenia. He postulated that those psychotic parents who succeeded in having children tended to have later onset and milder forms of illness so that the phenotypic 'depression' in parents in his study was the same illness genotypically as the schizophrenia in their offspring.

Most recent studies using strict and explicit diagnostic criteria, have found that there is no excess of schizophrenia among the relatives of patients with depression and although affective illness occurs in the relatives of schizophrenics, it is not more frequent than among the relatives of psychiatrically healthy controls (9,17). However, when the diagnosis of schizophrenic probands is broadened to include 'questionable' cases of schizophrenia, a small excess of depressive illness among relatives has been demonstrated (9). It has also been shown that although the family distribution of psychiatric illness is clearly different among the relatives of schizophrenic and depressed probands, this distinction is less marked for affective disorder probands who present with mania (7).

Interesting results on an associated theme have been produced by Odegaard (21) who found that about a third of cases of psychosis among relatives of schizophrenics were classified as affective psychosis if the proband had a disorder with prominent affective features or one resulting in 'no defect'. This compared with only one-sixth of psychotic relatives receiving an affective diagnosis where the index schizophrenic had a more severe and typical disorder. Odegaard remarked that conventional diagnoses in the major psychoses 'are helpful in the same way as the degrees of latitude and longitude for ocean navigation and not as unsurpassable walls erected between two parts of a population'. Not all researchers are comfortable with this dimensional approach. Those from a clinical background in particular have striven to retain diagnostic entities and it is convenient to summarize the arguments by switching to an equine metaphor.

Horses for Genetic Courses

Kendell (22) writing in the latest edition of the well known Edinburgh textbook suggests that if we consider schizophrenics as horses and manic depressives as donkeys then schizo-affectives may be:

(1) atypical horses
(2) atypical donkeys
(3) horses and donkeys yoked in pairs
(4) mules, or
(5) a completely distinct species; zebras

Kendell does not consider a sixth possibility, perhaps because at the outset it seems implausible. However, there has been a recent recurrence of interest in the notion of unitary psychosis and so we should, for completeness, postulate that the horses and donkeys may simply be different varieties of an 'Einheitspferd'. This is of course not just bad German but a bad hypothesis. The unitary psychosis hypothesis combines the worst of the categorical and the dimensional approaches to nosology. All varieties of psychoses become different manifestations of the same thing and there is no need to quantify or describe the variations. From the point of view of genetic analysis this beast takes us nowhere and the unitary psychosis hypothesis will not therefore be considered further here.

We should next consider whether schizo-affective disorder might be the chance occurrence of two psychoses, horses and donkeys yoked in pairs. The concept of co-morbidity has recently been much talked about especially in the United States where the application of DSM-III or similar criteria in epidemiological studies has resulted in many subjects fulfilling criteria for more than one diagnosis. The rules of a check-list classification procedure then take precedence over clinical common-sense and such individuals are deemed to have two or more disorders. Should we extend this concept and propose that schizo-affective illness is two conditions occurring simultaneously? The lifetime risk of schizophrenia based on calculations using the Camberwell Register is .86% (4). The lifetime risk of mania in Britain is also under 1% and an estimate, again based on the Camberwell Register, suggests that the morbid risk may be as low as .25% (23). Depression of sufficient severity to warrant inpatient treatment has a morbid risk of up to the age of 65 of 2.7%. Therefore the chance co-occurrence risk of schizophrenia and mania could be as low as 2 per 100,000 and the upper limit for the chance co-occurrence of schizophrenia and any severe affective disorder would be less than 3 per 10,000. But, as we know from clinical practice, an admixture of schizophrenic and affective symptoms is not a rarity. Indeed, in a recent consecutive series of 150 psychotic patients admitted to the acute general psychiatric wards of a London teaching hospital, approximately 20% fulfilled research diagnostic criteria for schizo-affective illness. We can therefore effectively exclude chance co-occurrence as the explanation of the great bulk of cases of schizo-affective disorder.

To choose between the remaining hypotheses requires us to look as closely as possible at the pedigrees of our subjects. Table 2 summarizes some recent studies of the families of schizo-affective probands. The family history study of Tsuang et al (25) suggested a familial association between schizo-affective disorder and affective illness but not schizophrenia. The study of Gershon et al (26) has been held to support this view. Although the morbid risk of affective disorder in Gershon's sample of schizo-affective patients is impressively high at 37%, there was also a small and almost significant (exact test p = .057) excess of schizophrenics among the relatives of schizo-affective probands (4%) compared with control relatives (0%). Gershon's results are therefore broadly in keeping with the other studies in Table 2 which, in general, show high risks of both schizophrenia and of affective illness in first degree relatives. It is also noteworthy that, with the exception of Gershon's study, all investigators find higher rates of schizophrenia or manic depression among relatives of schizo-affectives than of an intermediate form of psychosis.

Table 2   Psychotic illness in first degree relatives of
schizo-affective probands

Authors	schizophrenia	schizo-affective illness	affective illness
Shopsin et al (27)	8	0	13
Tsuang et al (25)	1	0	12
Angst et al (28)	11	0	35
Mendlewicz et al (29)	5	3	7
Scharfetter and Nusperli (30)	14	3	10
Gershon et al (26)	4	6	31
Kendler et al (31)	6	3	11

One major problem in comparing studies is that there is no
uniformity of diagnostic procedure. Investigators might in
some cases select predominantly affective schizo-affective
cases while others may choose predominantly schizophrenic
schizo-affectives. Some researchers have therefore attempted
to take this into account in their studies. Angst and his
colleagues (32) subdivided their probands but did not find
any marked differences in the patterns of illness among the
relatives of mainly schizophrenic or mainly affective
probands. By contrast, Kendler et al (31) in their
reassessment of the quaintly named 'Iowa non-500' series
found a more tidy pattern. Using the research diagnostic
criteria of Spitzer et al (14) these authors found that
mainly schizophrenic schizo-affective disorder had a close
familial relationship with schizophrenia but no apparent
relationship with affective illness. The remainder of the
schizo-affective sample was judged to be aetiologically
heterogenous showing possible familial relationships with
affective illness and remitting psychotic syndromes as well
as schizophrenia.

    Thus far, the familial evidence about schizo-affectives
suggests that some are horses, some are donkeys but rather
few are likely to be mules. However, a more direct test of
the mule hypothesis is to examine the offspring of so-called
dual matings of schizophrenics and manic depressives.
Unfortunately, the published evidence here is very scant.
Furthermore, the largest series of dual mating studies is now
old. Elsasser (33) reported 19 matings of schizophrenics
with manic depressive psychotics which produced a total of 68
offspring. Of these individuals, 6 developed schizophrenia
and 6 had manic depressive psychosis. Elsasser combined his
own data with previous reports in the literature and the
results are summarized in Table 3. As with the family data
discussed above, the Kraepelinian entities are commoner than
intermediate forms. The same pattern again emerges in the

most recent report of dual mating material from Denmark (34). Despite the inevitably small sample size the comparability of the pattern shown in Table 2 with that discovered by Elsasser 30 years earlier is striking. We must conclude that the mule hypothesis has little support. Among psychiatrically ill individuals at high genetic risk of both schizophrenia and manic depressive illness, an intermediate psychosis is the exception rather than the rule, most cases presenting as one typical psychosis or the other.

Table 3   Psychosis in the offspring dual matings
           (schizophrenia and manic depressive)

| | | | Morbid risk (%) | |
Authors	N	S	S-A	MD
Elsasser (33)	85	11	4	24
Fischer and Gottesman (34)	22	14	5	18

S = schizophrenia
SA = schizo-affective
MD = manic depressive

Only a minority of authors have taken the stance that there may be three (or more?) familial psychoses, the two major Kraepelinian forms and another unrelated disorder. This then is the zebra hypothesis and it suffers from the fact that various authors have used various methods to define and describe their syndrome. Thus cycloid psychosis (35), reactive psychosis (36) and atypical psychosis (37) have all been reported to show the tendency to 'breed true' within families. There have also been reports of large pedigrees in which multiple members have been affected by similar schizo-affective or atypical psychosis (38,39). Whether these non-Kraepelinian psychoses are similar entities or indeed whether they are entities at all is uncertain. Unfortunately, there have been no systematic twin or adoption studies to test whether the alleged familial consistency with these forms of atypical psychosis has a genetic rather than a culturally-transmitted basis. However, there is now a small but quite provocative literature based on sub-samples of twin series and isolated reports of twins and triplets.

Non-identical Psychosis in Identical Individuals

Cohen et al (40) reported on the relationship between schizo-affective psychosis, manic depression and schizophrenia in a twin sample obtained via the U.S. Veteran's Register. The pair-wise concordance was 7 out of 14 (50%) in MZ twins and DZ twins. However, there was an admixture of schizophrenic affective disorders and schizo-affective disorders among the MZ pairs concordant for psychosis. Bertelsen (41) has recently re-evaluated Scandinavian twin studies assessing twin pairs where one or other can be said to have a schizo-affective diagnosis. He concludes that there was high overall concordance for endogenous psychosis but only about one-third or MZ pairs were homotypical for schizo-affective disorder.

These findings suggest that schizo-affective illness is not a very pure category from a genetic viewpoint. However, until fairly recently, there have been no reports of a more dramatic type of discordance, that is of genetically identical individuals where one has typical schizophrenia and the other has typical manic depressive psychosis. However, accounts have now been forthcoming, the first of which was of non-identical psychosis in identical triplets (Table 4).

Two of the set had received hospital diagnoses of schizophrenia while the third, treated elsewhere, was concluded to have manic depressive illness. A re-evaluation of their illnesses using standardized methods and diagnoses by blind raters suggested that the discordance was not simply due to misdiagnosis or to a different diagnostic bias between clinicians at different hospitals. The triplets illustrate some of the short-comings of a strictly-applied Kraepelinian dichotomy but also warn us against an uncritical acceptance of operational criteria. Similar findings have since been reported in twins by Dalby et al (43) who applied DSM-III criteria. We have already mentioned, in a somewhat different context, the study of Farmer et al (13) where DSM-III criteria were applied to the Gottesman and Shields (11) schizophrenia twin series. Although DSM-III schizophrenia proved to be highly heritable, it was also found that 5 MZ schizophrenic probands had co-twins who could be assigned to DSM-III affective disorder categories. In 4 cases, the affectively ill co-twins could not be regarded as psychotic and so perhaps did not greatly embarrass the neo-Kraepelinian DSM-III dichotomy. However, the fifth was diagnosed as having major affective disorder with mood incongruent delusions. Farmer et al (13) therefore experimented with diagnostic groupings and investigated the effect of different combinations of DSM-III categories on MZ/DZ concordance ratios. Some of the results are summarized in Table 5.

It was found that using an expanded definition of 'schizophrenia' which included not just schizotypal personality but also atypical psychosis and affective disorder with mood incongruent delusions, the MZ/DZ concordance ratio increased. By contrast, broadening the diagnostic perspective much further to include any axis one diagnosis reduced the ratio. This suggests that the 'most genetic' definition of schizophrenia would be one which is rather broader than that embodied in DSM-III. These workers therefore concluded that although there was no wholly satisfactory solution to the apparent anomaly, it may be more reasonable, for the present, to classify 'mood incongruent' delusional patients as having schizophrenia. There is support for this view in the DSM-III family study of Kendler and colleagues (31) who found an excess of schizophrenia among the relatives of probands with mood incongruent psychotic affective illness and therefore argued that the

Table 4 Comparison of diagnoses on the "Z" triplets (McGuffin et al 1982)

| Triplet | Hospital diagnosis | Blind assessors | | | CATEGO* | RDC |
		First	Second	Third	NSM/DSMN	
G	Manic-depressive psychosis, manic type	Manic-depressive psychosis, manic type	Manic-depressive psychosis, manic type	Manic-depressive psychosis, manic type	NSM/DSMN	Episodes of mania and major depressive disorder (age 14-21). One epidose of schizo-affective psychosis, manic type (age 26).
R	Schizophrenia, unspecified	Schizophrenia, paranoid type	Schizophrenia, unspecified	Schizophrenia, schizoaffective type	NS+	Episodes of schizo affective psychosi manic and depresse type (age 15-20). Schizophrenia chronic course (age 20-).
M	Schizophrenia, unspecified	Schizophrenia, paranoid type	Schizophrenia, unspecified	Manic-depressive psychosis, ciruclar type	NS+	Episodes of schizo affective psychosi manic and depresse type (age 15-17). Schizophrenia, chronic course (age 17-).

RCD, Research Diagnostic Criteria (Spitzer et al 1975)
*CATEGO diagnosis at the X50 stage: NSMN/DSMN = nuclear syndrome plus mania/schizophrenia without first-rank symptoms plus mania;
NS+ = nuclear syndrome. The diagnosis of all three triplets resolved to S+ (schizophrenic) at the X9 stage

Table 5 Combinations of DSM - III Categories and the Effect on MZ/DZ Concordance Ratio

Data from Farmer et al (13)

Diagnoses	MZ Twins		DZ Twins		MZ/DZ
	No. of probands	Probandness concordance %	No. of proband	Probandness concordance %	Concordance ratio
Schizophrenia	21	47.6	21	9.5	5.01
Schizophrenia + schizotypal personality + atypical psychosis + affective disorder with mood incongruent delusions	22	59.1	26	7.7	7.68
schizophrenia plus any axis 1 diagnosis	27	70.4	34	29.4	2.39

presence of prominent affective features during the course of a chronic 'schizophrenia-like' illness need not be indicative of a familial predisposition to affective illness.

Models and Strategies

In attempting to construct models of the transmission of affective disorder which might take us further than Kendell's (22) attractive equine metaphor, we need to bear three points in mind. First, although they suffer many drawbacks, Kraepelin's two entities continue to provide the most serviceable working hypothesis that we have. Second, the separation between these two main entities is never as clean as we would wish and there is much in the accumulated genetic data which would favour a dimensional approach to classification. Third, however, we need categories both for clinical practice and genetic studies of the analysis of segregation.

The set of models which enable us to have our dimensional cake and slice it up, make use of the concepts of liability and threshold. Liability to develop a disorder (or group of disorders) is considered as a continuously distributed variable in the population. Only those individuals whose liability at some time exceeds a certain threshold manifest the disorder. It is possible to extend the concept to multiple thresholds so that, for example, a common milder form of disorder might result when liability exceeds a certain threshold while exceeding a second, more extreme threshold, produces a more severe, less common condition. In its original formulation, the liability threshold model was constructed under the assumption that the genetic contribution was polygenic. That is, multiple genes at different loci act in a predominantly additive fashion to contribute liability. If there are also multiple environmental factors again working in an additive fashion, the distribution of liability would tend to normality (44). However, the genetic component of liability might all, or in part, be due to a major gene. Under the general single major locus model (45,46), the major gene is a sole source of resemblance between relatives but under the more complicated mixed model (47) it is postulated that the genetic components of liability are due to a major gene acting on a polygenic background. Both the major gene and polygenes then contribute to resemblance between relatives. At present there is quite persuasive evidence (48) that a pure single major locus model is unsatisfactory in explaining the transmission of schizophrenia and so the likely choice is between polygenic or mixed model inheritance.

How then can the transmission of schizo-affective psychosis be accommodated by such models? Perhaps the most interesting attempt at model fitting was that carried out by Gershon and his colleagues (26). This group postulated that schizo-affective disorder lies on the same continuum of liability as affective disorder. They proposed that the commonest form of affective illness, unipolar depression, resulted when liability exceeded the first threshold. When liability exceeded a second more extreme threshold, bipolar disorder resulted while beyond the third, and most extreme threshold, lay schizo-affective disorder. The predictions of the model (what Reich and his colleagues (49) call the iso-correlational model) are complex, but essentially the expectation is that the proportion of affected relatives will be greatest where the probands have the most severe form of disorder, lowest where the probands have the commonest, mildest form of disorder, intermediate where the probands have the intermediate form of illness. This is indeed what Gershon and his colleagues found. The schizo-affectives, as we have seen in Table 1, had a very high proportion of first degree relatives with affective disorder. Nevertheless there was also, as we have earlier remarked, a nearly significant excess of schizophrenics among the relatives of schizo-affective probands so that Gershon's model although appealing and providing an acceptable statistical fit, is not wholly satisfactory. We are left with the strong suspicion that even in this sample, patients with schizo-affective disorder are a mixture of individuals with an affective liability, or a schizophrenic liability or, less commonly, a combination of both.

Many would argue that a purely statistical analysis can never provide us with a definitive answer. A potentially more attractive strategy is to search for linkage or association between varieties of psychosis and genetic markers. A genetic linkage strategy is fraught with difficulties when we are dealing with a disease trait where the mode of transmission is uncertain and so far studies of classical genetic markers in schizophrenia have yielded meagre positive findings (50).

More recently, the new genetics of recombinant DNA has resulted in a new generation of genetic markers, so called restriction fragment length polymorphisms (RFLP) which exist because of individual differences in DNA base sequences. Using RFLP's it has been possible to construct a virtually complete human genetic linkage map (51). This means that it is theoretically possible to perform a linkage study with a high chance of success of localizing genes of large effect for any inherited condition. Whether the strategy will work in complex conditions which are highly polygenic is less certain. However, great encouragement has come from the recent report of linkage between manic depressive illness and the H-ras locus on the short arm of chromosome 11 (52). Elsewhere, there have been new reports confirming earlier findings that some forms of manic depressive illness are transmitted as an X-linked dominant with incomplete penetrance (53,54). However, there have also been a number of reports of pedigrees where manic depressive illness is neither chromosome 11 linked (55) or X-linked (56). In the face of these discrepant findings, it is reasonable to conclude that manic depressive illness is an heterogenous condition existing in at least three forms, one X-linked, one chromosome 11 linked and another (or others) linked to neither marker locus.

Should we postulate that there is likely to be a similar degree of heterogeneity in other functional psychoses? This idea certainly seems plausible. However, one of the major difficulties in carrying out genetic research in psychiatry is that we are dealing with rather clumsy exophenotypes (11). That is we work purely with clinical signs and symptoms and so far have made no certain progress in developing more clear-cut endophenotypes which lie nearer to the primary gene products. In such circumstances, it is reasonable for us to suppose that the indistinct and over-lapping syndromes which we encounter may have multiple distinct aetiologies. However, the obverse of genetic heterogeneity is pleiotropy where a single major aetiological factor may have several apparently different manifestations. So far for schizophrenia as a whole, despite the persisting suspicion of heterogeneity, there is no hard evidence that multiple distinct subtypes exist (57). We must conclude that the same applies to schizophrenic illness in which there are prominent affective changes. The available genetic evidence suggests that some such patients are genotypically schizophrenic, some are genotypically affectively ill and in some the case remains 'not proven' either way. These are, however, interesting times when major strides are being made not only in molecular genetics but in other areas of high technology such as brain imaging applied to the study of neuropsychiatric conditions. As we head for the turn of the century, the prospects seem good that Kraepelin's major entities described at the last century's turn, will be better understood at an aetiological level and will cease to be regarded as mysterious and purely 'functional' conditions.

REFERENCES

1. Jaspers K: in General Psychopathology (trans. Hamilton MW, Hoeing J) Manchester, Manchester University Press, 1962.
2. American Psychiatric Association. Diagnostic and Statistical Manual for Mental Disorders. American Psychiatric Association, Washington, DC, 1980.
3. Brockington IF, Kendell RE, Leff JP: Definitions of schizophrenia: concordance and prediction of outcome. Psychol Med 7: 387-398, 1978.
4. Gottesman II, Shields J: Schizophrenia, The Epigenetic Puzzle, Cambridge, Cambridge University Press, 1982.
5. Abrams R, Taylor MA: The genetics of schizophrenia: a reassessment using modern criteria. Am J Psychiatry 140: 171-175, 1983.
6. Pope HG, Jonas J, Cohen BA et al: Heritability of schizophrenia. Am J Psychiatry 140: 132-133, 1983.
7. Tsuang MT, Winokur G, Crowe R: Morbidity risks of schizophrenia and affective disorders among first degree relatives of patients with schizophrenia, mania, depression and surgical conditions. Br J Psychiatry 137: 497-504, 1980.
8. Feighner JP, Robins E, Guze SB et al: Diagnostic criteria for use in psychiatric research. Arch Gen Psychiatry 26: 57-63, 1972.
9. Guze SB, Cloninger CR, Martin RL et al: A follow up and family study of schizophrenia. Arch Gen Psychiatry 40: 1273-1276, 1983.
10. Baron M, Gruen R, Kane J et al: Modern research criteria and the genetics of schizophrenia. Am J Psychiatry 142: 679-701, 1985.
11. Gottesman II, Shields J: Schizophrenia and Genetics: a Twin Study Vantage Point. London, Academic Press, 1972.

12. McGuffin P, Farmer AE, Gottesman II, et al: Twin concordance for operationally defined schizophrenia. Confirmation of familiality and heritability. Arch Gen Psychiatry 41: 541-545, 1984.

13. Farmer AE, McGuffin P, Gottesman II: Twin concordance for DSM-III schizophrenia: scrutinizing the validity of the definition. Arch Gen Psychiatry 44: 634-641, 1987.

14. Spitzer RL, Endicott J, Robins E: Research Diagnostic Criteria, Instrument no. 58. New York State Psychiatric Institute, New York, 1978.

15. Schneider K: Clinical Psychopathology. Translated by Hamilton MW, Grune and Stratton, London and New York, 1959.

16. Tienari P, Sorri A, Lahti J, et al: Interaction of genetic and psychosocial factors in schizophrenia. Acta Psychiatr Scand [supp.] 319: 19-30, 1985..

17. Kendler KS, Gruenberg AM: An independent antigen of the Danish adoption study of schizophrenia IV. The pattern of psychiatric illness as defined by DSM-II in adoptees and relatives. Arch Gen Psychiatry 41: 555-564, 1984.

18. Kety SS, Rosenthal D, Wender PH et al: Mental illness in the biological and adoptive families of individuals who have become schizophrenic. Behaviour Genetics 6: 219-225, 1976.

19. Kraepelin E: Manic Depressive Insanity and Paranoia. (Translated by Barclay RM) E & S Livingstone, Edinburgh 1922.

20. Slater E: Genetical causes of schizophrenia symptoms. Monatsschrift Fine Psychiatrie and Neurologie 113: 50-58, 1947 (reprinted in Man, Mind and Heredity, ed. Shields J, Gottesman II).

21. Odegaard O: The multifactorial theory of inheritance in predisposition to schizophrenia, in Genetic Factors in Schizophrenia, Edited by Kaplan AR, Thomas CC, Springfield, Illinois 256-275, 1972.

22. Kendell RE: Other functional psychoses, in Companion to Psychiatric Studies, 4th Ed., Edited by Kendell RE, Zealley AK, Churchill Livingston, Edinburgh, 1988.

23. Shields J: Unpublished data, 1977.

24. Stuart E, Kamakura N, Der G: How depressing life is - life-long morbidity risk for depressive disorder in the general population. J Affective Dis 7: 109-122, 1984.

25. Tsuang MT, Dempsey GM, Dvoredsky A et al: A family history of schizo-affective disorder. Biological Psychiatry 12: 331-338, 1977.

26. Gershon ES, Hamovit J, Guroff JJ et al: A family study of schizoaffective bipolar I, bipolar II, unipolar and normal control probands. Arch Gen Psychiatry 39: 1157-1167, 1982.

27. Shopsin B, Mendelweicz J, Suslac L et al: Genetics of affective disorders: II Morbidity risk and genetic transmission. Neuropsychobiology 2: 28-36, 1976.

28. Angst J, Felder W, Lohmeyer B: Schizoaffective psychoses heterogenous? Results of a genetic investigation. II. J Affective Dis 1: 155-165, 1979.

29. Mendlewicz J, Linkowski P, Guroff JJ et al: Color blindness linkage to bipolar manic depressive illness: new evidence. Arch Gen Psychiatry 36: 142-147, 1979.

30. Scharfetter C, Nusperli M: The group of schizophrenias, schizoaffective psychoses and affective disorders. Schizophr Bull 6: 586-591, 1980.

31. Kendler KS, Heath AC, Martin NG et al: Symptoms of anxiety and depression in a volunteer twin population. The aetiologic role of genetic and environmental factors. <u>Arch Gen Psychiatry</u> 43: 213-221, 1986.
32. Angst J, Felder W, Lohmeyer B: Schizo-affective disorders: Results of genetic investigation: I. <u>J Affective Dis</u> 1: 139-153, 1979.
33. Elsasser G: <u>Die Nachkommen Geisteskranker Elternpaare</u>. ed Thieme G, Stuttgart, 1952.
34. Fischer M, Gottesman II: A study of offspring of parents both hospitalized for psychiatric disorders, in <u>The Social Consequences of Psychiatric Illness</u>, Edited by Robins LN, Clayton PJ, Wing JR, Brunner/Mazel, New York, 1980.
35. Perris C: A study of cycloid psychosis. <u>Acta Psychiatr Scand Suppl</u> 253, 1974.
36. McCabe MS: Reactive psychoses. A clinical and genetic investigation. <u>Acta Psychiatrica Scandinavica Supplementum</u> 259, 1975.
37. Mitsuda H: <u>Clinical Genetics in Psychiatry</u>, Igaku Shoin, Tokyo, 1967.
38. Kaij L: Atypical endogenous psychosis. Report on a family. <u>Br J Psychiatry</u> 113: 415-422, 1967.
39. Walinder J: Recurrent familial psychosis of the schizo-affective type. <u>Acta Psychiatrica Scandinavica</u> 48: 274-283, 1972.
40. Cohen SM, Allen MG, Pollen W et al: Relationship of schizo-affective psychosis to manic-depressive psychosis and schizophrenia. Findings in 15, 909 veteran pairs. <u>Arch Gen Psychiatry</u> 26: 539-546, 1972.
41. Bertelsen A: Genetics studies in schizophrenia and schizoaffective disorders, in <u>A Genetic Perspective for Schizophrenic and Related Disorders</u>, Edited by Smeraldi E, Milan, 1988.
42. McGuffin P, Reveley A, Holland A: Identical triplets: non-identical psychosis? <u>Br J Psychiatry</u> 140: 1-6, 1982.
43. Dalby JT, Morgan D, Lee ML: Single case study. Schizophrenia and mania in identical twin brothers. <u>J Nerv Ment Dis</u> 174: 304-308, 1986.
44. Falconer DS: The inheritance of liability to certain disease, estimated from the incidence among relatives. <u>Annals of Human Genetics</u> 29: 51-76, 1965.
45. James J: Frequency in relatives for an all-or-none trait. <u>Annals of Human Genetics</u> 35: 47-49, 1971.
46. Reich T, James W, Morris CA: The use of multiple thresholds in determining the mode of transmission of semi-continuous traits. <u>Annals of Human Genetics</u> 36: 163-184, 1972.
47. Morton NE: <u>Outline of Genetic Epidemiology</u>, Basel, Karger, 1982.
48. O'Rourke DH, Gottesman II, Suarez BK et al: Refutation of the single locus model in the aetiology of schizophrenia. <u>Amer J Human Genetics</u> 33: 630-649, 1982.
49. Reich T, Cloninger CR, Wette R et al: The use of multiple thresholds and segregation analysis in analyzing the phenotypic heterogeneity of multifactorial traits. <u>Annals of Human Genetics</u> 42: 371, 1979.
50. McGuffin P, Stuart E: Genetic markers in schizophrenia. <u>Human Heredity</u> 36: 65-88, 1986.

51. Donis-Kellar H, Green P, Helms C et al: A genetic linkage map of the human genome. _Cell_ 51: 319-337, 1987.
52. Egeland JA, Gerhard DS, Pauls DL et al: Bipolar affective disorder linked to DNA markers on chromosome 11. _Nature_ 325: 783-787, 1987.
53. Mendlewicz J, Simon P, Sevy S et al: Polymorphic DNA marker and X chromosome and manic depression. _Lancet_ ii: 1230-1232, 1987.
54. Baron M, Risch N, Hamburger R et al: Genetic linkage between X chromosome markers and manic depression. _Nature_ 326: 806-808, 1987.
55. Hodgkinson S, Sherrington R, Gurling HAD et al: Molecular genetic evidence for heterogeneity in manic depression. _Nature_ 325: 805-806, 1987.
56. Gershon ES, Mendlewicz J, Gaspar M et al: A collaborative study of genetic linkage of bipolar manic depressive illness and red/green colour blindness. _Acta Psychiatr Scand_ 61: 319-338, 1980.
57. McGuffin P, Murray RM, Reveley AM: Genetic influences on the functional psychoses. _Br Med Bull_ 43: 531-556, 1987.

INTERHEMISPHERIC RELATIONSHIPS AND DEPRESSION IN

SCHIZOPHRENIA IN THE PERSPECTIVE OF CEREBRAL LATERALITY

Pierre Flor-Henry

Alberta Hospital Edmonton
Edmonton Alberta

In recent years there has been a greater understanding of the local circuits of regional cerebral disorganization, which increasingly, can be correlated with psychopathological manifestations or mental symptoms clusters. Since the brain of mammals, of primates, of homo sapiens is essentially a two brain system linked by a bridge, the corpus callosum, it is clear that lateralized perturbation of one hemisphere will necessarily have contralateral implications. Thus an understanding of the neurophysiology of the corpus callosum becomes essential. Norman Cook (1) provides an excellent and original review of the current knowledge on callosal transmission characteristics:

1. Most of the cortex except for the primary somatosensory the auditory and visual areas receive and send corpus callosal fibres.

2. The number of callosal fibres is of the order of 200 to 250 million.

3. There are approximately 125 million cortical columns per hemisphere, the fundamental functional unit of the cortex which when activated inhibit adjacent columnar units by surround inhibition.

4. Thus the approximate ratio of 2 callosal fibres for each cortical column implies that a topographical relationship between the hemispheres is anatomically feasible.

5. Because, in brain evolution, as the corpus callosum increases in size, there is a corresponding massive increase in functional brain asymmetry, the fundamental action of callosal transmission must be inhibitory rather than excitatory (because if it was excitatory with increasing efficiency of callosal transmission the two hemispheres would become more similar and more symmetrical. The opposite is, of course, the case).

6.   This is confirmed by physiological experiments which
      indicate that stimulation of the corpus callosum
      produces a brief excitation followed by prolonged
      inhibition at the termination of the callosal fibres,
      which originate and end, in areas 3 and 4 of the cortex.

7.   75% of callosal fibres form homotopic projections, the
      others are symmetrically heterotopic or project into the
      limbic system. Symmetrically heterotopic projections
      terminate in the homologous area contralaterally and
      ipsilaterally on one side.

8.   Subcortical brain stem ascending monoaminergic arousal
      pathways are bilaterally symmetrical in their
      neocortical projection.

9.   Thus in the above system cortical activation leads to a
      mirror image negative relationship between the cerebral
      hemispheres given the inhibitory function of the corpus
      callosum.

        It should be emphasized that contralateral inhibition
does not imply the absence of information transfer but leads
to the possibility of highly ordered superinordinate
constructs through the disinhibition of adjacent columns.
Another fundamental principle of brain organization which has
to be kept in mind, in attempting to understand
psychopathology-brain relationships, is that both in bird and
mammalian evolution, emotionality, aggression and sexual
arousal are three modalities which are dependent on neural
systems of the right hemisphere but are under regulatory
inhibition controlled from the left hemisphere (2). The
underlying organizational principle is of 1. Interhemispheric
activation; 2. Interhemispheric coupling and 3. Contralateral
inhibition (3). At the same time the right brain exercises a
feedback inhibitory regulation of the left brain as is
suggested by the remarkable experiments carried out by Renoux
et al (4) in France. These authors showed that the T
lymphocytic immuno efficiency in the rat depends on left
neocortical regions. When these are destroyed the T-cell
immunological functional capacity is destroyed. Further
there is atrophy of the thymus. With right neocortical
lesion there is no effect on the T-cell mediated
immunological system, which remains intact, but hypertrophy
of the thymus ensues. Renoux and Biziere conclude that the
intact left neocortex is essential for T-cell mediated
activity and depends on a balanced brain asymmetry in which
the right hemisphere controls the inductive influence of
signals emitted by the left hemisphere (5). These principles
fit remarkably well with the patterns of brain
disorganization seen in psychosis. I had been led to a
similar formulation of the relationship between the
hemispheres some ten years ago (6).

        If one accepts the postulate that the directionality of
disturbance of lateral hemispheric organization determines
the form of the consequent psychosis: disturbance of the left
hemisphere resulting in schizophrenic symptomatology; of the
right hemisphere resulting in depressive symptomatology, it
is clear that bilaterally asymmetrical lateral

disorganization, through disruption of callosal transfer, will produce contralateral effects. If one considers cognitive lateral organization in normals, it is well established that in dextral subjects of dextral stock, language functions are mediated by activation of the left hemisphere and visual spatial and affective functions are mediated by structures lateralized in the right hemisphere. Therefore, if in the psychoses, either schizophrenic, depressive, schizo-manic or schizo-depressive, there is bilateral asymmetrical disorganization, this should interfere with the cognitive lateral organization. As we wrote in Flor-Henry and Koles (7), "If it is accepted that normal cerebral functions require the simultaneous stabilizing of each hemisphere through transcallosal neural inhibition originating in the opposite hemisphere - and that in particular the maintenance of specific cognitive hemispheric specialization presupposes the contralateral inhibition of potentially rival specialized systems in the opposite side of the brain - then it is clear that the laterality hypothesis of the endogenous psychoses immediately account for these shifts. The extent to which the bilaterally asymmetrical hemispheric dysfunction (determining the psychosis) interferes with ipsilateral cognitive mechanisms - and the extent to which this process abolishes transcallosal neural inhibition, will at the same time be the basis for, and limit the degree of, (cognitive) laterality shifts possible in psychosis."

This has been confirmed by a number of observations in the intervening years. Hommes and Panhussyen (8) had, in an unusual and bold investigation, injected intracorotid amytal in dextral depressed patients who were neurologically normal and found, remarkably, that the more depressed the patients the less their left hemisphere was lateralized for language. There was a negative and significant inverse correlation between language lateralization in the left hemisphere and depth of depression. The implication is that, with deepening of depression correlated with increasing right hemisphere disorganization, there is contralateral inhibition of left hemispheric functions[1] (9,10,11). Stevens et al (12) showed with telemetric EEG studies of dextral schizophrenic patients that, when engaged in spatial tasks, instead of activating the right brain, they activated the left brain. Wexler and Heninger (13) with verbal dichotic stimuli in normal dextrals, found the expected superiority in detecting verbal signals in the right ear compared to the left ear. This effect was lost in all acute psychoses, depressive, manic, or schizophrenic, with the right ear advantage returning with symptomatic recovery. A contralateral left hemispheric disturbance is again suggested in psychosis. We noted in statistical EEG studies that manic patients appear to activate their right hemisphere during verbal task and depressed patients their left hemisphere during spatial tasks (7). Silberman et al (14) found that depressed patients show left visual field (or right hemispheric superiority) for the processing of verbal tasks, again a reversal of normal

_____

1A number of EEG studies confirm increasing changes in left hemispheric state significantly correlated with increasing depth of depression.

cognitive laterality. Davidson (15) found, in normals, that a stimulus tachistoscopically projected to the right hemisphere is perceived as unpleasant, whereas the same stimulus projected tachistoscopically to the left hemisphere is perceived as more pleasant. This is in line with the differential lateralization of different emotions, euphoric or positive emotions-left lateralized; dysphoric or negative emotions-right lateralized. In depressed patients this pattern was reversed, once more suggesting a shift of lateral organization in psychosis. Gaebel and Ulrich (16) examined alpha power characteristics in medicated schizophrenics at rest (eyes closed) and during visual motor tracking tasks. In both healthy controls and schizophrenics the performance of women was inferior to that of men. Patients with 'schizophrenia specific symptomatology' (i.e. with positive symptoms) were impaired on the visual tracking task whereas the 'withdrawal-retardation syndrome' had no impact on tracking performance. Whereas in normals performance on the visual task related to right hemisphere activation, in the schizophrenics it was associated with relative activation of the left side (17).

On the fundamental nature of psychosis, unitary or non-unitary, dimensional or categorical, it should be remembered that this question was present from the very first descriptions of dementia praecox and manic depressive insanity. Kraepelin (18), in his description of dementia praecox included a depression cluster with delusions, agitation, excitement, and catatonic stupor. Conversely in the Kraepelinian (19) definition of manic-depressive insanity: stupor and melancholia gravis with visual hallucinations, delusions, somatic delusions, auditory hallucinations were described. Thus, embedded in the heart of the manic-depressive insanity of Kraepelin, was a schizophrenic complex. Empirical work (20) on the symptomatology of psychosis, when not selecting typical forms, but randomly accepting all consecutive patients with psychosis for study, indicates that, with respect to depression and schizophrenia, there are no discreet dichotomies, no bimodal distributions of symptoms corresponding to two distinct psychoses, but a unimodal distribution, where at one extreme, are the rare "pure" schizophrenias and at the other the rare "pure" depressions, with mixed schizo-depressive forms constituting the majority. Some conclusions, relating to this issue, emerged from a neuropsychological study of psychosis which we undertook a few years ago (21). A series of unmedicated psychotics, (28 schizophrenics, 25 manics, 28 psychotic depressives) were studied with an extensive neuropsychological test battery. Globally, neuropsychological abnormality was found in 90% of schizophrenias, 80% in manias and 60% of psychotic depressions. There was a progressive increase in cerebral disorganization as measured by the neuropsychological parameters, least in depression, intermediate in mania and maximal in schizophrenia. Bilateral asymmetry of dysfunction, significantly, was left > right in schizophrenia and right > left in psychotic depressives and bilaterally symmetric in manics. A factor analysis produced seven factors which accounted for some 70% of the variance. The group means on this first factor, before rotation, which was the largest single factor on which most of the neuro-

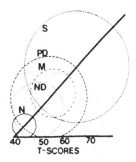

Group Severity

Figure 1   Means and standard deviations on profile level
            measure by group

psychological   variables   loaded,   (corresponding   to
severity of dysfunction) are shown in Figure 1.   This
emphasizes the extreme variability of the schizophrenics, in
spite of the fact that they were selected using the strictest
definition of schizophrenia available in schizophrenia
research (22). Variability, heterogeneity, is fundamental to
schizophrenia, it is not an accident because of the
arbitrariness or inaccuracy of diagnostic criteria.   It
should be noted that the psychotic depressives, also, have
large variability and the psychotic depressives who are
maximally impaired neuropsychologically, coincide, at the
level of severity of dysfunction, with the 50% of
schizophrenics who are least impaired. The manics are more
homogeneous and fall in an intermediate position between
depression and schizophrenia.   The continuous, overlapping
distribution of underlying neuropsychological structure in
the three psychoses is not compatible with a categorical, but
can only fit a dimensional representation.   In an earlier
neuropsychological study of psychoses (23) we had clinically
differentiated between schizophrenics, schizo-depressives and
schizo-manic patients. In the later study just summarized we

Table 1   Neurophychological Studies of Schizophrenic
           Psychoses

		N	Mean Age	Median Age
Schizophrenic	(S)	13	33	33
Schizodepressive	(SD)	8	39	34

WAIS (MEAN SCORES)

		Verbal	Performance	Full Scale
Schizophrenic	(S)	100	92	96
Schizodepressive	(SD)	95	89	92

33

Table 2   Tactual Performance Test Results

	Group	Median	Percent Impaired
Preferred Hand	S	441	60
	SD	341	28
NonPreferred Hand	S	390	50
	SD	349	71
Both Hands	S	223	70
	SD	180	28
Memory (Score)	S	7	30
	SD	5	57
Localization (Score)	S	2.5	60
	SD	3	57

Table 3   Purdue Pegboard Results

	Group	Median	Percent Impaired
Preferred Hand	S	14	8
	SD	13.5	37
NonPreferred Hand	S	13	16
	SD	12	50
Both Hands	S	11	25
	SD	10	62
Assemblies	S	27	75
	SD	25	62

Table 4 Miscellaneous Neuropsychological Test Results

	Group	Median	Percent Impaired
Oral Word Fluency	S	10.5	50
	SD	10.6	50
WILLIAMS VERBAL LEARNING (errors)	S	5	27
WILLIAMS NONVERBAL LEARNING (errors) Initial Position	S	1.5	30
	SD	6	71
WILLIAMS NONVERBAL LEARNING (errors) Rotated Position	S	0	30
	SD	9	85
SEASHORE RHYTHM TEST (errors)	S	4.5	30
	SD	7	57

Table 5 Motor Strength and Speed Results

	Group	Median	Percent Impaired
DYNAMOMETER Preferred Hand	S	29	33
	SD	27	37
NonPreferred Hand	S	21	33
	SD	26	50
FINGER TAPPING Preferred Hand	S	39	50
	SD	39	40
NonPreferred Hand	S	35	50
	SD	31	50

Table 6   Finger Localization Results

	Group	Median	Percent Impaired
FINGER LOCALIZATION (Single Stimulation) Preferred Hand	S	1	16
	SD	2	25
NonPreferred Hand	S	.5	25
	SD	2	37
FINGER LOCALIZATION (double stimulation) Preferred Hand	S	5	41
	SD	6	50
NonPreferred Hand	S	4.5	41
	SD	5	50

Table 7   Figertip Writing and Trail Making Test Results

	Group	Median	Percent Impaired
FINGERTIP WRITING Preferred Hand (errors)	S	1	20
	SD	2	28
FINGERTIP WRITING NonPreferred Hand (errors)	S	1	20
	SD	3	42
TRAIL MAKING TEST A	S	48	40
	SD	47	42
TRAIL MAKING TEST B	S	85	20
	SD	85	28

had selected typical forms, "pure" schizophrenia, "pure" mania and "pure" depression. A small sample of 13 "pure" schizophrenias and 8 schizo-depressives from the earlier investigation were reanalysed. The demographic and psychometric characteristics are shown on Table 1.

The neuropsychological characteristics of schizophrenia and schizo-depressive psychoses were expressed in terms of % with impairment, (% of subjects in each group showing deficit when compared to published age and sex matched norms). The neuropsychological characteristics of schizophrenics and schizodepressives are indicated in Tables 2-8.

It is apparent that, on a number of measures schizo-depressives have more right hemisphere dysfunction, the schizophrenics having more left hemisphere dysfunction (Tables 2, 3, 4 & 7). Green, (24) has carried out extensive studies with the monaural auditory comprehension test which he has developed. In general in schizophrenics 75% of patients have a left ear deficit and 25% have a right ear deficit in comprehension and immediate recall of stories heard monaurally in the left ear, and the right ear. The schizophrenics with a left ear deficit are acute positive symptomatology forms, while the right ear deficit occurs in chronic forms with negative, deficit symptomatology. In depression the left ear deficit is characteristic, in mania there is a tendency towards a right ear deficit. Importantly it has been found by Green, (and also by others) that in some schizophrenics with hallucinations refractory to neuroleptics, blocking with an earplug the ear with maximal deficit abolishes the auditory hallucinations. The interpretation of the cerebral repercussions of a left or right ear deficit is complex. The first assumption is that a left ear deficit implies a contralateral right hemisphere (temporal) locus of dysfunction, right ear deficit a contralateral left hemisphere (temporal) locus of dysfunction (Figure 4). Since verbal processing is mediated by the left hemisphere, the system is asymmetric, as verbal signals perceived by the left ear have to pass through the corpus callosum for processing. This is not true for right ear signals, which are directly projected to the left hemisphere. There is also evidence that efferent frontal-cochlear pathways are asymmetric. Lesions of the left frontal lobe produce ipsilateral ear extinction on dichotic tests but there is no corresponding pattern on the right side (25). Green's technique reveals a dichotomy in schizophrenia: left ear deficit-acute; right ear deficit-chronic. Other techniques also point to the dichotomous nature of schizophrenia. The directionality of electrodermal amplitude asymmetry, (26) indicates that acute schizophrenias with positive symptomatology have reduced electrodermal responsivity of the right hand compared to the left hand, whereas the chronic deficit forms with negative symptomatology have reduced electrodermal responsivity of the left hand compared to the right hand. The model proposed by Gruzelier (27) to explain the cerebral origins of electrodermal characteristic and asymmetry is that ipsilateral pathways are excitatory and contralateral pathways are inhibitory, for both hemispheres. However, the left hemisphere is inhibitory and the right hemisphere is excitatory (Figure 3). In such a system acute schizophrenia with electrodermal amplitude asymmetry of Right < Left directionality exhibits left hemisphere preponderance and fast habituation, correspondingly chronic schizophrenia is a state of relative right hemisphere preponderance, with slow habituation and EDA asymmetry Left < Right. Right hemi-

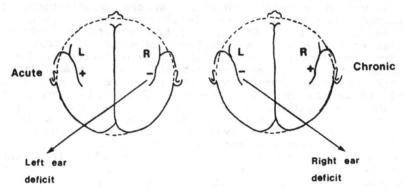

Figure 2.   Auditory comprehension Test in Schizophrenia

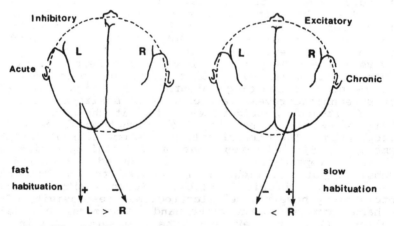

Figure 3.   Electrodermal Amplitude Asymmetry in Schizophrenia

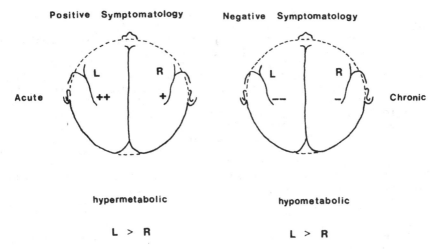

Figure 4. Positron Emission Tomography in Schizophrenia

spheric preponderance can be understood as a consequence of
dysfunction of the left hemisphere producing a contralateral
activation.   Conversely, in the acute psychoses, including
acute schizophrenia, the fundamental disturbance in the right
hemisphere produces, through callosal disinhibition, abnormal
contralateral   activation,   ie. left   preponderance.   Neuro-
metabolic studies (Positron Emission Tomography) in recent
years generally support these dichotomies since, from many
centres, (See Flor-Henry (28) for review), two different
types of abnormality of opposite directionality are described
in schizophrenia: hypermetabolic and hypometabolic. In both
the hypermetabolic and hypometabolic schizophrenias the
activity is bilaterally asymmetrical and the degree of
dysfunction, be it hypermetabolic or hypometabolic, is
greater in the left hemisphere.   The   acute   positive
symptomatology forms are bilaterally hypermetabolic,
temporal, left > right, whereas the chronic, negative
symptomatology forms are bilaterally hypometabolic, maximal
in the left frontal region (Figure 3).   Furthermore the
subcortical/cortical   activity   ratio   is   greater   in
schizophrenia, (because of greater relative subcortical
activity) than in normals and patients with more severe
psychopathology have higher left hemisphere activity (29). A
lateralized extension of this functional subcortical
asymmetry is seen in catatonia, characterized by marked
hypermetabolic activity of the left basal ganglia (30).
Machiyama (31) investigated monaural stimulation and CNS
excitability in schizophrenia by examining the auditory
evoked potentials (N100) to right and left ear monaural
stimulation.   He noted that in normal subjects right
hemisphere latencies are reduced on left ear stimulation,
which implies a greater degree of induced central inhibition.
In schizophrenia, left ear stimulation reduced parietal
latencies bilaterally but right ear stimulation reduced

parietal latencies bilaterally. Thus in schizophrenics the
state of central nervous system excitability changes in
opposite directions, depending on which ear is stimulated
monaurally: the left ear (increases neural inhibition), the
right ear (decreases it). In a complex manner these findings
may bear on the symptomatic improvement seen, occasionally,
in psychosis by blocking the ear with maximum deficit on the
Auditory Comprehension Test. Tomer and Flor-Henry (1988,
Biological Psychiatry in press) studied the effect of
neuroleptics on hemispheric functions in schizophrenic
patients, using the Mesulam cancelling test, which makes it
possible to determine which hemisphere is more efficient in
terms of visual analysis. It is found that schizophrenics
(N=15 patients, unmedicated) have a right field hemi-
inattention (i.e. left hemisphere dysfunction) in line with
expectations. As they improve on neuroleptics, however, they
develop left hemi-inattention, (right brain dysfunction) as
neuroleptics normalize left hemisphere functions. In a
functional sense, coinciding with symptomatic improvement, as
the left hemisphere normalizes, the right hemisphere
deteriorates. The effect is dose dependent, and is also
related to duration of neuroleptic exposure. A similar
implication follows from the work of Gaebel and Ulrich (32):
patients with more acute schizophrenic symptoms had greater
relative left hemispheric activation than those in remission
in the resting state. Patients who did not improve on
neuroleptics were in a state of relatively right hemispheric
activation compared to treatment responsive patients.
Another study which emphasizes neuroleptic - right
hemispheric interactions relating it to the depressive
complex in schizophrenia, is that of Koukkou et al (33) in
Zurich. She investigated paroxysmal EEG activity induced by
clozapine in 20 schizophrenics, and by haloperidol in another
19. The EEG was monitored on days 0, 3, 10, 20 and 30 after
initiation of medication, and changes were related to
baseline incidence on day 0. The EEG indices were correlated
with psychopathological syndrome scores: apathetic,
hallucinatory, hostile, manic, somatic-depressive, paranoid,
catatonic, and retarded depressive. Patients treated with
clozapine who had paroxysmal EEG changes received
significantly less clozapine than the patients who did not
show EEG changes; the former were significantly more improved
for the retarded depressive syndrome than the latter. There
was a negative correlation between frequency of EEG paroxysm
and intensity of depression, and a positive correlation
between frequency of EEG paroxysms and relief of this
depressive syndrome. Moreover, the frequency of the EEG
activity induced by clozapine was greater in the right than
in the left temporal region and the inverse relationship
between EEG paroxysms and intensity of depression was also
significantly higher in the right hemisphere than in the
left.

I will next discuss a single case, studied in detail, of
"Psychotic depression presenting as acute schizophrenia".
This was a man, age 26. The illness was of 6 to 12 months
duration with anxiety, auditory hallucinations: several
voices male and female commenting on his thoughts and
actions, mind influenced at a distance by his mother, and

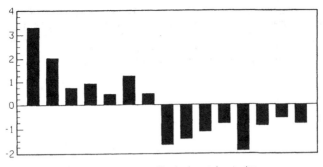

Figure 5. Neuropsychological Test Scores Pretreatment/Post
Treatment Standard Deviation Units

also by objects touching his head such as a hairbrush or
shampoos, experiences of thought deprivation and thought
insertion together with thought blocking, and olfactory
hallucinations were also present. He felt that a foul smell
emanated from his head. Subjectively his mood was empty,
objectively it was blunted. He had thoughts of suicide
intermittently but made no attempts. Sleep was disturbed, as
was his appetite, impairment of concentration was present and
there was loss of sexual drive. There had been one episode
four years earlier, of several months duration when he had
felt manic and full of energy, but this was not associated
with overspending, hyper-sexuality, grandiosity or sleep
disturbance and did not require treatment. One brother was
on Lithium. He was the second born of twins, delivered
eleven minutes after his brother, with a birth weight of 8
pounds, an easy child (the co-twin was hyperactive). He
completed high school successfully, by the age of 19 and he
had a good work record. He was single, heterosexual and
clinically presented as a case of classical schizophrenia.
After two months on neuroleptics, however, instead of
improving, the mental state worsened, he developed
extrapyramidal symptoms, which were asymmetrical and greater
on the left side of the body (implying that the illness was
fundamentally right hemispheric, i.e. in all probability a
mood psychosis). The neuroleptics were stopped and after two
induced seizures (ECT) and lithium the patient became
asymptomatic in two or three weeks. The neuropsychological
scores before and after treatment are shown in Table 8a. On
balance the scores are in the normal range before and after
treatment. However, if we examine the pattern of change
before and after treatment there is improvement for dominant
hemisphere functions and deterioration for most of the
variables relating to the right hemisphere (Table 8b).
The consequence of the treatment was a shift from relative
left brain dysfunction and good right brain efficiency before
treatment, to improved left and deteriorating right
hemispheric state after treatment and symptomatic recovery
(Figure 5). The clinical EEG before treatment showed left
frontal fast activity. The power spectral EEG (log of the
right/left power ratios) indicated a state of relative right
brain activation for all cerebral regions, except for the
temporal regions where a state of relative left brain

activation (corresponding to the left temporal fast activity on the clinical EEG) was seen. After treatment the left temporal fast activity disappeared, but bilateral frontal spike/wave discharges supervened. At no time (past or present) was there any evidence of epilepsy.

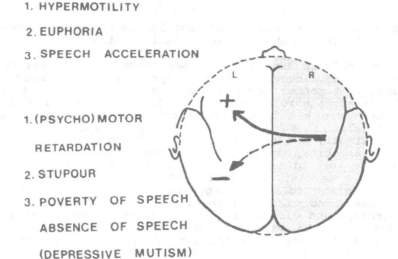

1. HYPERMOTILITY

2. EUPHORIA

3. SPEECH ACCELERATION

1. (PSYCHO) MOTOR

RETARDATION

2. STUPOUR

3. POVERTY OF SPEECH

ABSENCE OF SPEECH

(DEPRESSIVE MUTISM)

Figure 6. Induced Change From Contralateral Excitation (+) and Inhibition (-)

On the postulate that the fundamental locus of dysfunction in the affective psychoses is right hemispheric and that at a certain level of cerebral dysfunction there is a transcallosal effect which produces either contralateral excitation or contralateral inhibition, a model for the excited and/or depressed phases of the bipolar psychoses emerges. If contralateral (dominant) excitation supervenes leading to abnormal activation of left frontal systems this produces the symptoms of hypermotility, euphoria, speech acceleration, in other words the symptomatology of mania is triggered (induced activation of the left frontal zone from a systems). If the induced change is of contralateral retardation, stupor, poverty of speech, absence of speech,

Table 8a Neuropsychological Test and Retest Results

	Before Treatment	After Treatment
Oral Word Fluency	11.00	11.33
Written Word Fluency	11.33	12.33
Speech Sounds Perception	6.00	5.00
Williams Verbal Learning	8.00	1.00
Seashore Rhythm	5.00	8.00
Colored Progressive Matrices	1.00	.00
Memory-For-Designs	45.00	45.00
Trail Making Test A	23.50	29.70
Trail Making Test B	86.10	54.10
Symbol Digit Modalities:Written	50.00	43.00
Symbol Digit Modalities:Oral	56.00	57.00
Purdue Pegboard: Right Hand	15.00	14.00
Tactual Performance: Right Hand	353.60	284.40
Purdue Pegboard: Left Hand	14.00	11.00
Tactual Performance: Left Hand	152.90	174.30
Purdue Pegboard: Both Hands	12.00	10.00
Purdue Pegboard: Assemblies	38.00	36.00
Tactual Performance: Both Hands	108.00	81.20
Category Test	17.00	13.00
Wisconsin Card Sort: Errors	18.00	9.00
Wisconsin Card Sort: Subtests	6.00	6.00
Tactual Performance: Memory	8.00	9.00
Tactual Performance: Localization	5.00	4.00
Williams Clinical Memory:		
Nonverbal Learning: T1	2.0	.00
Nonverbal Learning: T2	.00	.00
Dynamometer: Right Hand	53.75	61.75
Finger Tapping: Right Hand	59.80	54.00
Name Writing: Right Hand	7.70	.49
Dynamometer: Left Hand	53.25	59.25
Finger Tapping: Left Hand	49.00	48.60
Name Writing: Left Hand	33.10	2.91
Finger Localization: Right Hand	3.00	4.00
L.J. Tactile Recognition: Right	13.30	27.10
Face-Hand Right Hand	.00	.00
Finger Localization: Left Hand	8.00	5.00
L.J. Tactile Recognition: Left	14.70	34.20
Face-Hand: Left Hand	.00	.00
L.J. Tactile Recognition: Both	12.60	16.70

WAIS-R	
Information	12.00
Digit Span	9.00
Vocabulary	17.00
Arithmetic	10.00
Comprehension	12.00
Similarities	15.00
Picture Completion	14.00
Picture Arrangement	13.00
Block Design	12.00
Object Assembly	12.00
Digit Symbol	8.00
Verbal IQ	113.00
Performance IQ	113.00
Full Scale IQ	115.00

Table 8b Neuropsychology Test Scores Changes Expressed In
Standing Deviation Units After Lithium Treatment
For Case 12022 (Retest 12119)

NEUROPSYCHOLOGY TEST IMPROVEMENTS	Standard Deviations	
Williams Verbal Learning	+ 3.30	
Trail Making B	+ 2.03	
Tactual Performance - Right Hand	+ .75	Dominant
Dynamometer - Right Hand	+ .93	5
Speech Sounds Perception	+ .47	
Finger Localization - Left hand	+ 1.25	Nondominant
Tactual performance - Memory	+ .48	2

NEUROPSYCHOLOGY TEST DECREMENTS		
Purdue Pegboard - Left Hand	- 1.65	
Purdue Pegboard - Both Hand	- 1.41	
Namewriting - Left Hand	- 1.13	Nondominant
Trail Making A	- .80	6
Seashore Rhythm	- 1.90	
Symbol Digit - Written	- .88	
Purdue Pegboard - Right hand	- .56	Dominant
Finger Tapping - Right Hand	- .80	2

depressive stupor, (again from a right brain origin (Figure
6). With an extension of the contralateral excitation to
involve the left temporal region, the schizo-manias, thought
disordered manias, acute schizophrenias emerge, corresponding
to the left preponderance electrodermal asymmetry dichotomy,
primary disruption of right brain modulation of left brain
inhibition, the opposite state is evoked: psychomotor
the left ear deficit dichotomy, the (Left > Right)
hypermetabolic asymmetric states. Depression in
schizophrenia can occur as a result of two different
processes: from a primary right hemispheric basis as in the
"acute", or "atypical" schizophrenias, or, alternatively, in
the "true" schizophrenias, dementia praecox forms, where the
fundamental defect is of the dominant hemisphere, from a left
hemispheric basis with consequent contralateral disinhibition
disrupting the emotionality systems characteristic of the
right hemisphere.

REFERENCES

1. Cook ND: The Brain Code. Mechanisms of Information
    Transfer and the Role of the Corpus Callosum. London,
    New York, Methuen & Co Ltd, 1986.
2. Denenberg VH: Hemispheric laterality in animals and the
    effects of early experience. Behavioral Sciences 4: 1-
    21, 1981.
3. Denenberg VH: General systems theory, brain
    organization and early experiences. Am J Physiology 238:
    5-13, 1980.
4. Renoux G, Biziere K, Renoux M, Guillaumin J-M, Degenne
    D: A balanced brain asymmetry modulates T cell-mediated
    events. J Neuroimmunology 5: 227-238, 1983.

5. Renoux G, Biziere K: Brain neocortex lateralized control of immune recognition. Integrative Psychiatry 4: 32-40, 1986.

6. Flor-Henry P: The endogenous psychoses: a reflection of lateralized dysfunction of the anterior limbic system. In Limbic Mechanisms, Edited by KE Livingston and O hornykiewica, New York, Plenum Publishing Corporation, 1978.

7. Flor-Henry P, Koles ZJ: EEG studies in depression, mania and normals: evidence for partial shifts of laterality in the affective psychoses. Advances in Biolog Psychiatry 4: 21-43, 1980.

8. Hommes OR, Panhuyssen LHHM: Bilateral intracarotid amytal injection. Psych Neurololgy Neurochirurgia 73: 447-459, 1970.

9. d'Elia G, Perris C: Cerebral functional dominance and depression. Acta Psychiat Scand 49: 191-197, 1973.

10. Nystrom C, Matousek M, Hallstrom T: Relationships between EEG and clinical characteristics in major depressive disorder. Acta Psychiat Scand 73: 390-394, 1986.

11. Cazard P, Pollak V, Jouvent R, Leboyer M, Grob R, Lesevre N: Hemisphere asymmetry of alpha burst sequential organization in depression. Int J Psychophysiology (In Press) 1988.

12. Stevens JR, Bigelow L, Denney D, Lipkin J, Livermore AH, Rauscher F, Wyatt RJ: Telemetered EEG-EOG during psychotic behaviors of schizophrenia. Arch Gen Psychiatry 36: 251-262, 1979.

13. Wexler BE, heninger GR: Alterations in cerebral laterality during acute psychotic illness. Arch Gen Psychiatry 36: 278-284, 1979.

14. Silberman EK, Weingartner H, Stillman R, Chen HJ, Post RM: Altered lateralization of cognitive processes in depressed women. Am J Psychiatry 140: 1340-1344, 1983.

15. Davidson RJ: Cerebral asymmetry and the nature of emotion: implications for the study of individual differences and psychopathology. In Cerebral Dynamics, Laterality and Psychopathology, Edited by R. Takahashi, P. Flor-Henry, J. Bruzelier, S-I Niwa, Amsterdam, Elsevier Science Publishers 1987.

16. Gaebel W, Ulrich G: Visuomotor performance and alpha-power topography in the EEG: syndrome relationships in schizophrenia: in , Cerebral Dynamics, Laterality and Psychopathology Edited by R. Takahashi, P. Flor-Henry, J. Bruzelier, S-I Niea, Elsevier, Amsterdam, 1987.

17. Gaebel W, Ulrich G: Visuomotor performance of schizophrenic patients and normal controls. II. Results of a visual search task. Pharmacopsychiatry 19: 190-191, 1986.

18. Kraepelin E: Dementia praecox and paraphrenia. In Textbook of Psychiatry, (Barclay RM (Translator) Robertson GM (ed)) Vol iii, Edinburgh, E & S Livingstone, 1919.

19. Kraepelin E: Manic-Depressive Insanity and Paranoia Edinburgh, E.S. Livingstone, 1921.

20. Kendell RE, Brockington IF: The identification of disease entities and the relationship between schizophrenic and affective psychoses. Br J Psychiatry 137: 324-331, 1980.

21. Flor-Henry P, Fromm-Auch D, Schopflocher D:
    Neuropsychological dimensions in psychopathology. In
    Laterality and Psychopathology Edited by P. Flor-Henry,
    J. Gruzelier, Amsterdam, Elsevier Science Publishers,
    1983.
22. Taylor MA, Abarams R, Gaztanaga P: Manic-depressive
    illness and schizophrenia: a partial validation of
    research diagnostic criteria utilizing
    neuropsychological testing. Comp Psychiatry 16: 91-96,
    1975.
23. Flor-Henry P, Yeudall LT: Neuropsychological
    investigation of schizophrenia and manic-depressive
    psychoses. In Hemispheric Asymmetries of Function in
    Psychopathology Edited by J. Gruzelier and P. Flor-
    Henry Amsterdam, Elsevier/North Holland Biomedical
    Press, 1979.
24. Green P: Interference between the two ears in speech
    comprehension and the effect of an earplug in
    psychiatric and cerebral-lesioned patients. In Cerebral
    Dynamics, Laterality and Psychopathology Edited by R.
    Takahashi, P. Flor-Henry, J. Gruzelier, S-I Niwa
    Amsterdam, Elsevier Science Publishers, 287-298, 1987.
25. Ogden JA: Ipsilateral auditory extinction following
    frontal and basal ganglia lesions of the left
    hemisphere. Neuropsychologia 23: 143-159, 1985.
26. Gruzelier JH: Hemispheric imbalances masquerading as
    paranoid and non-paranoid syndromes? Schizophr Bull
    7(4): 662-673, 1981.
27. Gruzelier JH, Eves FF, Connolly JF: Habituation and
    phasic reactivity in the electrodermal system:
    Reciprocal hemispheric influences. Physiological
    Psychology 9: 313-317, 1981.
28. Flor-Henry P: Cerebral dynamics, laterality and
    psychopathology: a commentary. In Cerebral Dynamics,
    Laterality and Psychopathology, Edited by R. Takahashi,
    P. Flor-Henry, J. Gruzelier, S-I Niwa, Amsterdam,
    Elsevier Science Publishers, 3-21, 1987.
29. Gur RE: Regional brain dysfunction in schizophrenia:
    PET and regional cerebral blood flow studies. In
    Cerebral Dynamics, Laterality and Psychopathology Edited
    by R. Takahashi, P. Flor-Henry, J. Gruzelier, S-I Niwa,
    Amsterdam, Elsevier Science Publishers, 503-512, 1987.
30. Luchins DJ, Metz J, Marks R, Cooper MB: Basal ganglia
    regional glucose metabolism asymmetry during catatonic
    episodes. Presented at Annual Meeting of the American
    Society of Biological Psychiatry, Montreal, June
    Abstract #159, 219 of program, 1988.
31. Machiyama Y, Shiihara Y, Kubota F: Topograhpic and
    temporal aspects of information processing abnormalities
    in schizophrenia. In Cerebral Dynamics, Laterality and
    Psychopathology Edited by R. Takahashi, P. Flor-Henry,
    J. Gruzelier, S-I Niwa, Amsterdam, Elsevier Science
    Publishers, 211-220, 1987.
32. Gaebel W, Ulrich G: Topographical distribution of
    absolute alpha-power in schizophrenic outpatients - On-
    drug responders vs. Nonresponders. Pharmacopsychiatry
    19: 222-223, 1986.

Section 2

PHENOMENOLOGY

INTRODUCTION

Donald Addington

This section presents six papers which address the issue
of the phenomenology of depression in schizophrenia. For
many schizophrenic patients depression is undeniably a
serious problem. Many questions about the nature and course
of depression in schizophrenia are largely unanswered. Is
depression a reaction to the psychotic episode, is it a
concomitant of it, or is it a result of antipsychotic
medication? Do clinicians unavoidably confuse depression in
a schizophrenic with the residual schizophrenic defect state?
The recent interest in the concept of negative symptoms adds
to the complexity. Furthermore, in order to delineate the
boundary between the negative symptom syndrome and depressive
symptoms, measures of negative symptoms need to be more
clearly defined. It is issues such as these that the authors
of the following papers address.

Sampson, Young and Tsuang attempt to define the symptoms
that differentiate negative or deficit state schizophrenia
from other schizophrenias and from affective illnesses. This
group of researchers from Harvard compared a sample of
schizophrenics with a sample of unipolar depressives, many of
whom had psychotic features. It was found that, although
schizophrenics had significantly more negative symptoms than
the depressed patients, there was an overlap in negative
symptoms present between the two groups, such that these two
groups could not be discriminated on the basis of negative
symptoms. Discrimination was only apparent when patients
with depressed or manic mood were eliminated from the group
of schizophrenic patients.

Thus, these authors conclude that the negative symptom
syndrome is found in patients without depressed or manic
mood. Furthermore, they suggest that there is a continuum of
severity of SANS symptoms, on which schizophrenics with
affective disturbance lie between schizophrenics without
affective disturbance mood and depressed unipolar patients.

In their Canadian study, Addington and Addington were
interested in comparing, in a sample of 50 schizophrenics,
three measures of depression, taken first at the acute stage
of the illness and again six months later at a time of
relative remission in addition, the overlap of depressive
symptoms with ratings of negative symptoms, assessed by the
SANS, was examined.

Results indicated that the different measures of
depression overlapped with the SANS ratings and that the

overlap was partly attributable to attentional impairment. Furthermore, the measures of depression were more highly correlated at follow-up than to the acute stage, depression in a period of relative remission is more typical of the depression observed in patients with a depressive disorder.

Owens and Johnstone, from the Northwick Park Hospital in England, compare symptoms of depression, negative symptoms and extra-pyramidal symptoms among a number of diagnostic groups which are obtained from a consecutive series of admissions with functional psychosis. These patients were also participating in a medication study and thus, measures of depression, negative symptoms, and extrapyramidal symptoms were made before patients were allocated a formal diagnosis.

From their wide array of rich clinical data, these authors report that their schizophrenic subtypes all showed significant but similar levels of depressive symptoms which were significantly less than the psychotic depression group. Levels of depression differed quantitatively but not qualitively between groups. Using a narrow definition of negative symptoms, they found no evidence of an overlap between extrapyramidal symptoms, depression, and negative symptoms.

The paper by Romney attempts to discriminate the difference between schizophrenia and psychotic depression. He traces the development of the hypothesis that paranoid schizophrenia and psychotic depression may be construed as lying on a bipolar continuum with the purest form of each condition located at either end. This author presents evidence to support this hypothesis but acknowledges that such evidence is largely circumstantial. In concluding, Romney describes the beginning of research designed to test this hypothesis.

In their paper, Green, Nuechterlein, Mintz, and Ventura from The University of California, Los Angeles present interesting methods of data analyses on a uniquely detailed longitudinal data set of schizophrenics. These researchers are interested in identifying whether prospective ratings of depression and psychotic symptoms across multiple episodes would allow one to determine whether there is a temporal relationship between depressive symptoms and the onset of the psychotic episode.

Their use of the analyses adds to an understanding of the temporal course and relationship of psychotic symptoms. The analyses show that the onset of depressive periods occurred with a significantly higher frequency concurrent with psychotic onset, but did not differ from expected frequencies from any other time period.

Michael Pogue-Geile, from the University of Pittsburgh, reports on an investigation designed to examine the nature of the relationship between depression and negative symptoms in a sample of 44 stabilized outpatients with schizophrenia or schizoaffective disorder. Results of this study, which used a modified form of the SANS, suggested that narrowly defined negative symptoms are statistically independent of self-report measures of depression.

The various findings and conclusions of this series of papers have important implications for researchers examining the causes and treatment of depression in schizophrenia. In these papers the need for careful selection of specific subscale measures in order to separate the syndromes of depression, negative symptoms, positive symptoms, and extrapyramidal symptoms has been emphasized. In particular, the identification of a core negative syndrome is essential in order to make progress in identifying its genetic and neurobiological correlates.

NEGATIVE SYMPTOMS IN SCHIZOPHRENIA:

DEPRESSIVE OR DEFICIT SYNDROME

Jacqueline A. Samson, Alexander Young and
Ming T. Tsuang

Harvard Medical School Boston Massachusetts and
Brockton\West Roxbury VA Medical Center
Brockton Massachussets

The development of a comprehensive understanding of the etiology, treatment, and prevention of psychiatric disorders has been slow and hampered by uncertainties about the existence of etiologically distinct illnesses within the larger domain of psychiatric illness. Early attempts at subtyping psychiatric disorders relied heavily on clinical observation and the abilities of clinicians to recognize patterns of recurring symptoms and outcome within their clinics and practices. From this approach emerged the seminal descriptions of Kraepelin and Bleuler, and the foundation of current clinical nosology. Key to the early systems (and to our current DSM III-R system) (1) was the distinction between affective illnesses conceptualized as disorders of mood disturbance (such as manic-depressive psychosis), and schizophrenic illnesses conceptualized as being associated with a fundamental defect in the emotions or volition [according to Kraepelin (2)] or capacity for appropriate affective response or goal-directed activity [(according to Bleuler (3)]. Schizophrenic illnesses were seen as more permanent in nature, with a progressively deteriorating course, and often characterized by disordered thinking, delusional beliefs, or impairments in perceptual processes (hallucinations).

Interestingly, the distinction between schizophrenic and affective illnesses drawn by early nosologists relied less on symptom specificity than on longitudinal course and premorbid history. Hence, similar symptoms might be manifested by patients suffering from very different disease processes, so that interpretation of the entire grouping of signs and symptoms (or the syndrome) was necessary before diagnosis. This issue of symptom nonspecificity was particularly salient with regard to the interpretation of affect in schizophrenia versus manic-depressive illnesses. In Kraepelin's description of dementia praecox, loss of capacity for affect and restricted affective expression were included as characteristic of several of his subtypes. Affective illnesses, however, were thought to be associated with aberrations in mood states, such that there was limited capacity to express or experience the <u>full range</u> of appropriate emotions with restriction to a single predominant (often intense) mood, or alternation between two dominant moods (depressed and euphoric). Kraepelin described it this way (2) (pages 264-265):

"while in dementia praecox the emotions are silent, without the patients noticing the disorder, or being disturbed about it, the manic-depressive patients complain in despairing accents of the feeling of inward desolation and emptiness, of their inability to feel joy or sorrow, although in their conduct emotional reactions of great vivacity appear".

Included in Kraepelin's early descriptions were also detailed observations of the external behaviors to be considered in making a differential diagnosis between subtypes of schizophrenic disorders and affective illness (2) (page 265).

"Very important is the distinction of negativism from anxious resistance and the inhibition of will in manic-depressive insanity. Even the behavior at the approach and greeting of the physician permits certain conclusions to be made. The negativistic (dementia praecox) patient does not look up, hides himself perhaps, turns away or stares straight in front of him, and does not betray by any movement of a muscle that he is aware of anything. All the same he usually perceives better than the manic-depressive patient, who indeed also perhaps remains mute and motionless, but still in his glance, in the expression of his face, in slight attempts at movement, acceleration of the pulse, flushing, stoppage of respiration, lets it be seen that he has felt the impressions".

Subtleties such as those described by Kraepelin require a practiced clinical eye as well as the opportunity for prolonged contact with the patient. Rarely do researchers concerned with subtyping have the luxury of either. As a consequence, research definitions of blunted or flat affect in schizophrenia tend to rely on gross signs such as decreased eye contact, increased latency of speech, decreased amount of speech, decreased movements and decreased gestures.

Further differentiating current nosology from earlier works, are the types of symptoms included as criteria for schizophrenic illness. As outlined by Kety (4), current diagnostic descriptions rely heavily on the presence of delusions, hallucinations, and thought disorder. Affective state is considered only as an associated symptom or as a component of criteria for subtype assignment. For instance, in DSM III-R, the presence of flat or inappropriate affect is listed as an exclusion for paranoid subtype and is one of several inclusion criteria for disorganized subtype. Hence, patients who are labelled as schizophrenic in the DSM system are identified primarily on the basis of the severity of "positive" symptomatology, and may or may not also exhibit the volitional and emotional deficits described by Kraepelin.

Negative Syndrome in Schizophrenia

Crow (5) has recently stimulated renewed interest in the pursuit of a definition of a subtype of schizophrenia characterized by "core defect" or negative symptoms such as affective flattening, poverty of speech and loss of drive. Crow contends that schizophrenic patients whose primary symptom picture is consistent with these negative symptoms show poor outcome, poor response to neuroleptics, and structural changes in the brain. Since the initial hypothesis, some support has emerged to validate the "defect state" or "Type II" schizophrenia (6) although not all studies concur (7).

Crow's work dovetails with increased interest in the prognostic significance of negative symptoms by American researchers, in particular the work of Strauss (8-10). Carpenter (11-12) and Andreasen (13-15). Much of the excitement in the research arena of negative symptoms concerns the identification of a core group of signs and symptoms that may comprise a unique disease entity or subtype (see 15). Clinical descriptions (5,6,12) of defect or negative syndrome schizophrenia acknowledge that symptom stability, as well as course of illness, neuroleptic drug use and coexisting positive symptomatology must be considered in addition to the nature of the symptom presentation. However, research definitions, in the attempt to operationalize fuzzy clinical concepts and to reduce idiosyncratic clinical biases, rely more strictly on the assessment of specific symptoms and subsequent classifications by clusters of symptoms (or syndrome) to assign patients to groupings. And, as more investigators enter this endeavor, and more symptom scales are developed to assess different sets of theoretically "negative" symptoms, the diversity in definition of negative syndrome increases (for a discussion see 16-17).

Depressive Symptoms in Schizophrenia

The appearance of depressive features in the context of a schizophrenic illness is a well documented phenomenon (18). Depressive features may precede the first appearance of psychiatric symptoms (19) may co-exist with psychotic features (20-24), or may appear first in the context of a post-psychotic phase of illness (25-27).

Important to differential diagnosis is the duration and severity of depressive features, as well as the prominence of the depressive features relative to psychotic symptomatology. Clinically, differential affective diagnoses are made on the basis of the presence or absence of the full syndrome of depressive illness. Although syndromal definitions differ across diagnostic systems, the common approach is first to establish the presence of a persistent and unremitting depressive mood (or anhedonia in the DSM III systems), and then to define the full syndrome on the basis of associated features such as cognitive impairment, somatic or vegetative signs. In patients who present with symptoms and signs that meet full syndrome criteria for both schizophrenia and major depressive disorder, a diagnosis of schizoaffective disorder is assigned. If the symptoms of schizophrenia clearly predominate over the affective symptoms, a diagnosis of schizophrenia is assigned. If the affective features predominate, the diagnosis becomes major depressive disorder, with psychotic features. Hence, the depressive syndrome is commonly observed in psychotic patients and differential diagnosis is contingent on the course of illness and the relative ratio of severity of psychotic to affective features.

Long-term outcome studies of schizoaffective illness have generally found that affective patients do better over time than schizophrenic patients and that schizoaffective patients show outcome intermediate to that of schizophrenic and affective illnesses. Hence, in schizophrenic illnesses, the coexistence of the depressive syndrome has become associated with more favorable prognosis, leading some to see the group of schizoaffective illnesses as "good-prognosis" schizophrenia (31-36). However, not all studies have been in agreement (see 29). But what of the patients who present with predominant features of schizophrenic illness, but who also show less severe symptoms (or fluctuating symptoms) of depressive disorder? Where does the boundary exist that delineates those schizophrenic patients with a more favorable prognosis from the remainder of the sample?

Negative Symptoms and Depressive Symptoms in Schizophrenia

The previous review discussed the parameters of two purported subtypes of schizophrenia; schizophrenia with negative syndrome and schizophrenia with depressive syndrome. Negative syndrome schizophrenia has been associated with a chronic deteriorating course and poor long-term outcome, whereas depressive syndrome in schizophrenia has been associated with good prognosis and a remitting course. Clearly, the literature predicts very different outcomes for these 2 subtypes, yet the boundary between symptoms characteristic of each remains indistinct (see 17). Associated features of depressive illnesses in DSM III-R include, anhedonia, psychomotor retardation, anergia, decreased concentration or decreased decisiveness. Variations of these features are also included in the definitions of negative syndrome put forth by Andreasen (14), Abrams and Taylor (37), Iager et al (38) Kay et al (39).

Studies that specifically examine negative and depressive symptoms in schizophrenic patients have been few. Using scales derived from the Brief Psychiatric Rating Scale, Lindenmayer et al (40) found that depressive features correlated with negative symptoms at baseline assessment and each predicted favorable outcome at 2 years post-baseline in a group of young schizophrenic patients. When the depressive factor was partialled out of the statistical analyses, 59% of the variance shared between negative symptoms at baseline and 2 year outcome ratings was lost. However, at the 2 year follow-up, depressive and negative symptoms were not related.

Lindenmayer et al (40-41) point out that the prognostic significance of negative symptoms may be phase specific, with the greatest prediction of poor outcome in older, more chronic patients, and a greater overlap with depressive symptoms in younger and more acutely ill patients.

Using a different research strategy, Pogue-Geile and Harrow (42) reported that negative symptoms tended not to co-occur with self-reports of depression in young schizophrenic patients and went on to show that the total severity of negative symptoms was higher in their schizophrenic sample than in a comparison sample of unipolar depressive patients at an 18 month post-baseline follow-up study.

While the samples studied by Lindenmayer et al (40) and Pogue-Geile and Harrow (42) were very similar (young, acute schizophrenics), the symptoms used to define the negative syndrome varied. Pogue-Geile and Harrow restricted their analyses to symptom measures of flat affect, poverty of speech and psychomotor retardation (rated from the Present State Examination). In contrast, Lindenmayer et al defined negative syndrome by symptoms of blunted affect, emotional withdrawal, poor rapport, passive/apathetic social withdrawal, difficulty in abstract reasoning, lack of spontaneity and flow of conversation, and stereotyped thinking. Symptoms were rated from the Brief Psychiatric Rating scale, using the subscale methods of Kay et al (41). Moreover, Lindenmayer et al defined depressive syndrome in their schizophrenic patients by summing the symptom item scores for depressed mood, guilt and motor retardation. Pogue-Geile and Harrow utilized a control group of patients meeting Research Diagnostic Criteria for unipolar major depressive disorder.

PRELIMINARY STUDIES

We have recently initiated a preliminary series of studies as a step toward empirically defining the core symptoms that differentiate negative or deficit state schizophrenia from other schizophrenias and affective illnesses (43). In this work, we examined scores on the Schedule for the Assessment of Negative Symptoms (SANS) (44) in combination with affective subscales derived from the Schedule for Affective Disorders and Schizophrenia (SADS) interview (45). Data were drawn from an initial sampling of patients (all male) participating in the Brockton VA - Harvard 1000 study.

In view of the widespread disagreement about the definition and conceptualization of the "atypical psychoses" and schizoaffective disorders, the Brockton VA - Harvard 1000 study was broadly designed to include all patients who present with psychotic symptoms that do not meet DSM III-R criteria for schizophrenia or affective disorder. The study is being conducted at the Brockton campus of the Brockton-West Roxbury Veterans Administration Medical Center (hence the Brockton VA). The Brockton-West Roxbury VA Medical Center is an affiliate of the Harvard Medical School, and personnel for the project are drawn from both institutions. The research design calls for baseline and follow-up assessments of a core group of 400 "atypical psychotic" or schizoaffective patients and patient control groups of 200 schizophrenic patients, 200 bipolar manic-depressive patients and 200 unipolar major depressive disorder patients. Thus the larger study includes prospective longitudinal evaluations of 1000 patients.

As part of the protocol of the Brockton study, patients are interviewed using the SADS instrument and SANS ratings are assigned by the research interviewer on the basis of the SADS interview. Interviews average from 4 to 6 hours in duration and are distributed over a 2-3 day period for each patient.

To ensure accuracy of diagnostic assignment, symptoms assessed during the self-report SADS interview are cross-validated by reports of treating clinicians and by chart reviews. This material is then presented to a panel of 2 psychiatrists and 2 psychologists who review all sources of information and assign a consensus diagnosis according to a number of diagnostic systems.

In the first of the negative symptom studies, we selected from our initial pool of subjects only those patients who met full DSM III-R criteria for schizophrenia or unipolar major depressive disorder. The resulting sample contained 55 schizophrenics and 15 unipolar depressed patients. Most of the sample was receiving neuroleptic medication at the time of the study (53% of the depressive and 94% of the schizophrenic patients). The fact that such a large percentage of our unipolar sample were on neuroleptic medication reflects the nature of our sample. A good portion of our unipolar depressed patients evidenced some psychotic features (30%) at the time of interview. This circumstance strengthened our study, in that the effects of neuroleptic medication on negative symptoms were distributed across both of our comparison groups.

In keeping with the approach adopted by Pogue-Geile and Harrow (42) as well as Kay et al (41), data pertaining to items from the attention subscale of the SANS were not included in our measure of negative symptoms. Instead, we focussed on SANS items contained in subscales for affective flattening, alogia, anergia, and anhedonia.

We were first interested in knowing whether the SANS items, as a whole, differentiated the unipolar depressive from the schizophrenic sample. To arrive at a total severity scale for negative symptoms, item scores (but not the global scores) were summed from the affective flattening, alogia, anergia and anhedonia sections of the SANS. These total SANS severity scores were then compared across the DSM III-R schizophrenic and unipolar depressed patient groups. The mean ($\pm$SD) SANS severity score for the unipolar patients was 10.5 ± 8.0, which was significantly lower (F = 6.45, p < .01) than that of the schizophrenic patient group (mean $\pm$ SD = 19.4 ± 12.9).

Since we were particularly interested in the actual symptoms that were unique to the schizophrenic group, we went on to perform a multiple regression analysis (Statistical Analysis System, GLM procedure) to predict diagnostic group on the basis of a "best fit" equation combining scores from specific negative symptoms. This procedure had the advantage of controlling for the interrelationship among the 21 SANS items, and allowed the identification of those symptoms that contributed most to the discrimination of schizophrenic from depressed patients.

As is shown in Table 1, using the DSM III-R definition of schizophrenia to define patient groups resulted in SANS items accounting for 44% of the variance in diagnosis, which did not meet statistical significance levels.

Table 1   Prediction of Diagnostic Group by SANS

	Uni. vs. Schiz. Total	Uni.vs. Schiz. No Dep.	Uni.vs. Schiz. No Dep. No Manic
Total R-Squared	.44	.69	.86
Total F	1.28	2.31*	
Symptom Predictors: Type III F Values			
Decreased Facial Expression	2.48	6.72*	9.73**
Inappropriate Affect	4.09*	7.85**	9.04**
Increased Latency	.38	3.37+	3.38+
Anergia	2.47	6.79*	4.30+
Decreased Intimacy	1.62	4.68	3.77+
Decreased Friendships	.04	3.33+	4.50+

+ P < .10
* P < .05
**P < .01

It was clear from these analyses (and from visual inspection of the data points) that, although our schizophrenic sample showed an overall greater severity of negative symptoms than our unipolar depressed sample, there was a great deal of overlap in negative symptom presentation between the two patient groups. Moreover, our GLM procedures showed that the two patient groups could not be statistically discriminated on the basis of negative symptom presentation. Since many of our schizophrenic patients showed coexisting, but less prominent signs of depressive disorder, we considered the possibility that mild depressive features within some of schizophrenic sample might be obscuring more obvious differences in negative symptomatology between our non-depressive schizophrenics and our unipolar depressed patients. Because the core symptom of depressive illness is depressive mood, and because many of the associated symptoms used to define depressive syndrome are also used to define negative syndrome in schizophrenia, we elected to classify depressive illness only on the basis of the component of depressive mood. To classify patients, we defined severity of depressive mood on the basis of patient ratings on the depressive mood subscale of the SADS, and then assigned our schizophrenic patients into two groups on the basis of depressive mood subscale scores. To arrive at a reasonable cut score for separating the patient groups we examined the distribution of SADS mood subscale scores for the schizophrenic group as a whole. There was a break in the distribution pattern at the severity score of 10, and this score was subsequently used for classification. This resulted in the identification of 18 schizophrenic patients with depressive mood and 37 schizophrenics without depressive mood.

Negative symptom analyses were then repeated, this time restricting our contrasting patient groupings to those schizophrenics with no depressed mood and unipolar depressed patients. This resulted (see Table 1) in a significant prediction of diagnostic group membership by SANS items ($F = 2.31$, $p < .05$) and multiple R of .83, suggesting that 69% of the variance in diagnosis was predicted by SANS item scores. In particular, schizophrenic patients showed more inappropriate affect ($F = 7.58$, $p < .01$) decreased facial depression ($F = 6.72$, $p < .05$), and anergia ($F = 4.68$, $p < .05$).

Encouraged by these results, we continued on to question whether there were other patients included in our schizophrenic sample who showed signs of an affective component to the illness, but in the form of manic or hypomanic mood. (Although bipolar manic-depressives were not included in our depressive patient sample, DSM III-R diagnostic labels did not provide a means of separating out from our schizophrenic sample patients with bipolar-like features of mild severity). To accomplish this we again turned to the SADS interview, and selected items reflecting euphoric, irritable or expansive mood or grandiosity. Item scores were summed to devise a subscale score to assign schizophrenic patients into two groups. Again, visual inspection methods were used to identify a reasonable out-score for classification. This resulted in the identification of 7 schizophrenic patients with significant manic mood.

As the final step, we redefined our patient comparison groups so that only those schizophrenic patients without significant signs of depressed mood or manic mood were included in the comparison with unipolar depressed patients. As is shown in Table 1 this approach resulted in a significant prediction to group membership ($F = 9.79$, $p < .01$) with 86% of the variance in diagnoses explained by negative symptoms. The symptoms showing the greatest discriminatory power included facial expression ($F = 9.79$, $p < .01$) and inappropriate affect ($F = 9.04$, $p < .01$).

The total SANS severity scores for the schizophrenic sample with depressed or manic mood fell midway between those of other schizophrenics and those of unipolar depressives.

CONCLUSIONS

Results from our preliminary analysis suggested that negative symptoms were generally more severe in the schizophrenic patients than in the unipolar patients. These findings are consistent with the report of Pogue-Geile and Harrow (42). However, an examination of the distribution of negative symptoms severity scores indicated a good amount of overlap between the two patients groupings, particularly at the mild to moderate levels of severity.

The multiple regression analysis used to predict group membership on the basis of specific negative symptoms showed that significant discrimination of schizophrenic and unipolar depressed patient groupings was only achieved after those schizophrenics with a significant component of depressive mood were eliminated from the larger group of schizophrenic patients. Significance levels of the predictive equation improved even further when schizophrenic patients with manic mood were also eliminated.

Since our schizophrenic sample was based on the DSM-III-R definitions, there were no patients included who met criteria for true schizoaffective disorder or for major depressive disorder superimposed on residual schizophrenia. Hence, the patients who were eliminated from the negative symptom analysis in these studies were predominantly schizophrenic in presentation with some associated features of affective illness. This fact is important for two reasons. First, our symptom studies suggest that there are a good number of schizophrenic patients who present with concomitant mood disorder, but who do not meet DSM-III-R criteria for schizoaffective illness. These patients are not synonymous with "negative syndrome" schizophrenics, and are primarily delineated by the presence of depressive or manic mood. While these patients do not present with severe scores for "negative symptoms", they do show mild to moderate scores for SANS severity.

The second point to be made is that these findings suggest that schizophrenic patients with symptoms of depressive or manic mood show mean SANS symptom severity scores that are midway between those shown by the unipolar depressed patients and those shown by the schizophrenic patients without depressive or manic mood. Hence, it appears that these patient groupings reflect two continua of affective symptomatology and negative symptomatology. The patients with the most clear affective diagnoses show the lowest SANS scores, the patients with schizophrenic diagnoses who show concomitant mood disorder show intermediate SANS scores, and schizophrenic patients without concomitant mood disorders show the most severe SANS scores. A multidimensional and continuous model of psychopathology would best fit our symptom findings.

The implications of these results for the establishment of negative syndrome schizophrenia are that researchers concerned with symptom measurement must be extremely cautious in their selection of symptom items and patient groupings to be used in analyses. Our findings imply that the negative syndrome schizophrenia patient group is found among schizophrenic patients who do not show signs of depressed or manic mood. Hence, eliminating these patients may be the first step in elucidating the syndrome (see Figure 1).

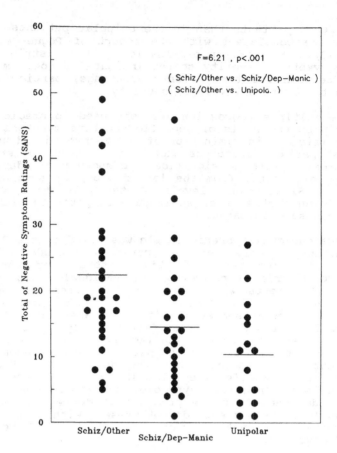

Figure 1. SANS Total: Schizophrenics with Depressed or Manic Mood

ACKNOWLEDGEMENTS

This work was funded in part by a grant from the Veterans Administration to Dr. Ming T. Tsuang. The authors wish to thank Dr. John Simpson for helpful comments and Mrs. Cindy Dion for preparation of the manuscript.

REFERENCES

1.   American Psychiatric Association: Diagnostic and Statistical Manual of Mental Disorders, Third Edition, Revised. Washington, DC, American Psychiatric Association, 1987.

2. Kraepelin E: Dementia Praecox and Paraphrenia. Translated by Barclay RM, edited by Robertson GM. Huntington, New York, Robert E. Krieger Publishing Co. Inc, 1971.
3. Bleuler E: Dementia Praecox or The Group of Schizophrenias. Translated by Zinkin J. New York, International Universities Press, 1950.
4. Kety SS: The concept of schizophrenia, in Controversies in Schizophrenia. Edited by Alpert M. New York, Guilford Press, 1985.
5. Crow TJ: The molecular pathology of schizophrenia: More than one disease process? Br Med J 280: 66-68, 1980.
6. Crow TJ: The two-syndrome concept: Origins and current status. Schizophr Bull 11: 471-485, 1985.
7. Goldberg S: Negative and deficit symptoms in Schizophrenia do respond to neuroleptics. Schizophr Bull 11: 453-456, 1985.
8. Strauss JS, Carpenter WT, Bartko JJ: The diagnosis and understanding of Schizophrenia: III. Speculations on the processes that underlie Schizophrrenic symptoms and signs. Schizophr Bull (Experimental issue No. 11): 61-69, 1974.
9. Gift TE, Strauss JS, Kokes RF et al: Schizophrenia: affect and outcome. Am J Psychiatry 137: 580-585, 1980.
10. Strauss JS: Negative symptoms: future developments of the concept. Schizophr Bull 11: 457-460, 1985.
11. Carpenter WT, Heinrichs DW, Alphs LD: Treatment of negative symptoms. Schizophr Bull 11: 440-452, 1985.
12. Carpenter WT, Heinrichs DW, Wagman AMI: Deficit and nondeficit forms of Schizophrenia: the concept. Am J Psychiatry 145: 578-582, 1988.
13. Andreasen NC: Affective flattening and the criteria for Schizophrenia. Am J Psychiatry 136: 944-947, 1979.
14. Andreasen NC: Negative symptoms in Schizophrenia, definition and reliability. Arch Gen Psychiatry 39: 784-788, 1982.
15. Andreasen NC: Positive vs negative Schizophrenia: a critical evaluation. Schizophr Bull 11: 380-389, 1985.
16. Lewine RJ, Sommers AA: Clinical definition of negative symptoms as a reflection of theory and methodology in Controversies in Schizophrenia. Edited by Alpert M. New York, Guilford Press, 1985.
17. Sommers AA: "Negative symptoms": conceptual and methodological problems. Schizophr Bull 11: 364-378, 1985.
18. Grinker RR: Anhedonia and depression in Schizophrrenia, in Depression and Human Existence. Edited by Anthony EJ, Benedek T, Boston, Little, Brown and Co, 1975.
19. Cooper JE: Diagnostic change in a longitudinal study of psychiatric patients. Br J Psychiatry 113: 129-142, 1967.
20. Bowers MB, Astrachan BA: Depression in acute schizophrenic psychoses. Am J Psychiatry 123: 976-979, 1967.
21. Shanfield S, Tucker GJ, Harrow M et al: The schizophrenic patient and depressive symptomatology. J Nerv Men Dis 151: 203-210, 1970.
22. McGlashan TH, Carpenter WT: Affective symptoms and the diagnosis of schizophrenia. Schizophr Bull 5: 547-553, 1979.

23. Brockington IF, Kendell RE, Wainwright S: Depressed patients with schizophrenic or paranoid symptoms. Psychological Medicine 10: 665-675, 1980.
24. Moller H, vonZerssen D: Depressive states occurring during the neuroleptic treatment of schizophrenia. Schizophr Bull 8: 109-117, 1982.
25. Vaillant GE: The natural history of the remitting schizophrenias. Am J Psychiatry 120: 367-376, 1963.
26. McGlashan T, Carpenter WT: Post-psychotic depression in schizophrenia. Arch Gen Psychiatry 33: 231-239, 1976.
27. Johnson DAW: The significance of depression in the prediction of relapse in chronic schizophrenia. Br J Psychiatry 152: 320-323, 1988.
28. Tsuang MT, Simpson JC: Schizoaffective disorder: concept and reality. Schizophr Bull 10: 14-25, 1984.
29. Harrow M, Grossman L: Outcome in schizoaffective disorders: a critical review and re-evaluation of the literature. Schizophr Bull 10: 87-108, 1984.
30. Samson JA, Simpson JC, Tsuang MT: Outcome studies of schizoaffective disorders. Schizophr Bull, in press.
31. Vaillant GE: A historical review of the remitting schizophrenics. J Nerv Ment Dis 138: 48-56, 1964.
32. Stephens JH, Astrup C, Mangrum JC: Prognostic factors in recovered and deteriorating schizophrenics. Am J Psychiatry 122: 116-121, 1966.
33. McCabe MS, Fowler RC, Cadoret RJ et al: Symptom differences in schizophrenia with good and poor prognosis. Am J Psychiatry 128: 1239-1243, 1972.
34. Taylor MA, Abrams R: Manic-depressive illness and good prognosis schizophrenia. Am J Psychiatry 132: 741-742, 1975.
35. Knight RA, Roff JD, Barnett J, Moss JL: Concurrent and predictive validity of thought disorder and affectivity: a 22-year follow-up of acute schizophrenics. J Abn Psychol 88: 1-12, 1979.
36. Tsuang MT, Dempsey GM: Long-term outcome of major psychoses II. Schizoaffective disorders compared with schizophrenia, affective disorders and a surgical control group. Arch Gen Psychiatry 36: 1302-1304, 1979.
37. Abrams R, Taylor MA: A rating scale for emotional blunting. Am J Psychiatry 135: 226-229, 1978.
38. Iager AC, Kirch DG, Wyatt RG: A negative symptom rating scale. Psychiatry Res 16: 27-36, 1985.
39. Kay SR, Fishzbein A, Opler LA: The positive and negative syndrome scale (PANSS) for schizophrenia. Schizophr Bull 13: 261-275, 1987.
40. Lindenmayer JP, Kay SR, Friedman C: Negative and positive schizophrenic syndromes after the acute phase: a prospective follow-up. Comp Psychiatry 27: 276-286, 1986.
41. Kay SR, Fiszbein A, Lindenmayer JP et al: Positive and negative syndromes in schizophrenia as a function of chronicity. Acta Psychiatr Scand 74: 507-518, 1986.
42. Pogue-Geile MF, Harrow M: Negative and positive symptoms in schizophrenia and depression: a follow-up. Schizophr Bull 10: 371-387, 1984.
43. Samson JA, Young A, Tsuang M et al: Negative symptoms in schizophrenia and depression. In New Research Abstracts, American Psychiatric Association Annual Meetings. American Psychiatric Association, 1988.

44. Andreasen NC: Scale for the assessment of negative symptoms. Department of Psychiatry, College of Medicine, The University of Iowa, Iowa City, Iowa, 1984.
45. Endicott J, Spitzer RL: A diagnostic interview: The Schedule for Affective Disorders and Schizophrenia (SADS). <u>Arch Gen Psychiatry</u> 35: 837-844, 1978.

THE ASSESSMENT OF DEPRESSION IN SCHIZOPHRENIA

Donald Addington and Jean Addington

Department of Psychiatry, Foothills Hospital and
University of Calgary and Department of Psychology
Holy Cross Hospital, Calgary Alberta

The presence of depressive symptoms and syndromes in schizophrenia were described by Kraepelin (1) and Bleuler (2). Subsequently, much of the attention devoted to depression in schizophrenia focused on its significance for prognosis and diagnosis (3-8).

The recent increase in attention to depression in schizophrenia has emphasized its importance for morbidity (9,10) and mortality (11,13) and the need for its treatment. This in turn has led to a requirement for better methods of assessing depression in schizophrenia, which is complicated by a number of factors. These include the temporal variability of depressive syndromes (14,15), the poor correlation between self report and observer reports of depression (16,18), and the overlap with extrapyramidal syndromes (19-20) and negative symptoms (21).

A variety of measures have been used to assess depression in schizophrenia: a scale derived from the Psychiatric Assessment Interview (22); different sets of syndromes from the P.S.E. (23,24); the Raskin Scale (25); the Hamilton Depression Rating Scale (26,27); and the Carroll Rating Scale (28). More recently, the DSM III definition was used to assess the prevalence of depression in chronic schizophrenia (29); to investigate DST non suppression in depressed chronic schizophrenics (30); and to select schizophrenics for antidepressant medication (31).

Furthermore, it has been shown (32), that depression rating scales used with acutely ill schizophrenics, show high total scale reliability but lower intra-item reliability. This suggests that such scales may measure different dimensions in acutely ill schizophrenics than they do in patients with affective disorders. In assessing depression in schizophrenics, it is apparent that a variety of measures have been used with little attention paid to their comparability or validity. While some authors have measured only depressive symptoms, (22-28) others have assessed a full depressive syndrome (29-31).

In light of the difficulty in comparing studies using different measure of depression and the potential overlap between depression and negative symptoms, this study was designed to compare three measures of depression at two stages of the schizophrenic illness and to assess their overlap with the negative syndrome.

METHOD

The sample consisted of 50 consecutive admissions with schizophrenia to two psychiatric admission units in a general teaching hospital. Subjects were required to have a diagnosis of schizophrenia according to DSM III criteria. Diagnosis was made on the basis of data combined from interviews with a modified form of the PSE (32) chart review and family interview. Patients were all competent to give informed consent and were all receiving antipsychotic medication. Subjects were excluded if they had any of the following: a) evidence of an organic central nervous system disorder; b) significant or habitual drug or alcohol abuse in the past year; c) mental retardation; or d) Carroll's DST exclusion criteria (33).

All patients completed a data sheet which was designed by the investigators and used to obtain necessary demographic and background information.

Subjects were assessed on a number of measures, the Scale for the Assessment of Negative Symptoms (SANS) (34). The Scale for Assessment of Positive Symptoms (34) and the Hamilton Depression Rating Scale (HDRS) (35). Interrater reliability was established by the principal investigator on a separate series of videotaped interviews at the Clinical Research Center at UCLA. Interrater reliability for individual PSE items was 83% and for the SANS subscales and total score rating 80%.

RESULTS

The mean age of the sample was 30.9 years. The mean time since first hospitalization was 8.3 years, mean number of admissions 5.2 mean age at first admission 22.4 years.

A total of 50 patients met study criteria and 41 were successfully assessed at a 6 month followup. Comparison between patients' symptom ratings at Time 1 and Time 2 are presented in Table I indicating a significant decline in positive and negative symptoms but a less clear cut decline in depressive symptoms.

However, the frequency of major depressive episodes did not differ significantly from time 1 to time 2, 24% were depressed at time 1 while only 17% were depressed at time 2.

The overlap between negative symptoms and depression was explored in a number of ways. Table 2 presents the intercorrelations of measures of depression and negative symptoms at Time 1 and Time 2. At time 1 correlations between the scale measures were higher than correlations between the scales and the presence of a major depressive episode. This difference did not exist at time 2. There were significant relationships between negative symptoms and all measures of depression at both time 1 and time 2.

Table 1   Symptom Comparison's Time 1 vs. Time 2

	MEAN TIME I	MEAN TIME II	T VALUE	2TAIL PROBABILITY
NEGATIVE SYMPTOMS	36.6	26.8	5.48	0.000**
POSITIVE SYMPTOMS	33.5	10.3	9.54	0.000**
HAMILTON (HDRS)	12.9	10.0	2.32	0.025
PSE(D)	6.5	3.3	5.08	0.000**

*   P < 0.01

**   P < 0.001

Relationship between depression and individual negative symptoms are further explored by multiple regression analysis Table III. This shows that MDE was significantly predicted by attentional impairment, alogia, avolition and anhedonia but not by affective flattening.

Discriminant function analysis Table III shows that at time I both PSE(D) and HDRS predicted the presence of MDE but HDRS did not add any further predictive power. The combination correctly classified 80% of grouped cases. At time 2 table IV both measures significantly contributed to the variance and together correctly classified 95% of grouped cases.

Table 2. Intercorrelations of Measures of Depression and Negative Symptoms

Time 1 and Time 2

	MDE		HDRS		PSED	
	Time I	Time II	Time I	Time II	Time I	Time II
Major Depressive Episode MDE	1	1				
Hamilton Depression Rating Scale HDRS	0.43*	0.72*	1	1		
Present State Examination Depression PSED	0.50**	0.71**	0.63**	0.69**	1	
Scale for Assessment of Negative Symptoms SANS	0.37*	0.32*	0.46*	0.42*	0.52**	0.45**

Table 3. Results of Stepwise Multiple Regression Analysis of
Negative Symptoms on Major Depressive Episode

Time I

SIGNIFICANT PREDICTOR VARIABLES	SIMPLE R	R^2	CHANGE IN R^2	BETA	DEGREE OF FREEDOM	STEPDOWN F
Attentional Impairment	0.37	0.135	0.135	0.477	48.1	7.48**
Alogia	0.24	0.167	0.032	0.480	47.2	4.72*
Avolition/ Apathy	0.34	0.178	0.011	0.540	46.3	3.33*
Anhedonia	0.19	0.188	0.010	0.261	45.4	2.61*
Affective Flattening	0.16	0.189	0.000	0.614	44.5	2.05

* $P < 0.05$
** $P < 0.01$

Table 4   Discriminant Function Analysis and Classification
           Results at Time 1

DEPENDENT VARIABLE MAJOR DEPRESSIVE EPISODE

INDEPENDENT Variable	WILKS LAMBDA	SIG	RAO'S	SIG	CHANGE IN V	SIG
PSE(D)	0.715	0.00	19.17	0.00	19.17	0.00
HDRS	0.699	0.00	20.70	0.00	1.54	0.21

CLASSIFICATION RESULTS

ACTUAL GROUP	NO. OF CASES	PREDICTED NOT DEPRESSED	PREDICTED DEPRESSED
Not Depressed	38	29 (76.3%)	9 (23.7%)
Depressed	12	1 (8.3%)	11 (91.7%)

PERCENT OF GROUPED CASES CORRECTLY CLASSIFIED 80%.

Table 5   Discriminant Function Analysis and Classification
          Results at Time 2

DEPENDENT VARIABLE MAJOR DEPRESSIVE EPISODE

Independent Variable	Wilks LAMBDA	SIG	RAO'S	SIG	CHANGE V	SIG
HDRS	0.477	0.000	42.6	0.000	42.6	0.000
PSE(D)	0.393	0.000	60.1	0.000	17.5	0.000

CLASSIFICATION RESULTS

ACTUAL GROUP	NO. OF CASES	PREDICTED NOT DEPRESSED	PREDICTED DEPRESSED
Not Depressed	34	32 (94.1%)	2 (5.9%)
Depressed	7	0 (0.0%)	7 (100%)

PERCENT OF GROUPED CASES CORRECTLY CLASSIFIED 95.12%

CONCLUSIONS

The sample of patients described in this study is fairly typical for an acute care facility. There was a range of duration of illness and a range of outcomes.

This present paper has two major findings. Firstly, the present data indicates that there is an overlap between the measure of negative symptoms and the measures of depression used in the study. Secondly, analysis of correlation, regression and discrimination suggest that the different measures of depression are more congruent with each other at the follow-up stage, than at time 1.

Thus, at time 2 the depression reported is more typical of that found in patients suffering from depression. If as this study suggests, different measures of depression appear to be measuring different things in the acute phase of schizophrenia, then it is not surprising that it is difficult to establish criteria for schizoaffective disorder.

Finally, this study has implications for the assessment of negative symptoms of schizophrenia. The significant change in negative symptoms between the two assessments indicates that the scale does not reflect an enduring stable entity. The Affective Flattening scale was the one scale showing no correlation with depressive symptoms and suggests that a narrower definition placing emphasis on affective flattening would better reflect the presence of an enduring negative or deficit syndrome (37).

REFERENCES

1. Kraepelin E: Dementia Praecox and Paraphrenia. Translated by R. Mary Barclay, Robert E. Kriegler Publishing Co., Huntington, New York, pp 103-106, 1971.
2. Bleuler E: Dementia Praecox or the group of Schizophrenias. translated by Zinkin J, International Universities Press, new York, pp 304-305, 1950.
3. Vaillant GE: A historical review of the remitting schizophrenias. J Nerv Ment Dis 138: 48-56, 1965.
4. Strauss JS, Carpenter WT: Prediction of outcome in schizophrenia III. Five year outcome and its predictions. Arch Gen Psychiatry 34: 159-163, 1977.
5. Gopelrud E, Depue R: The diagnostic ambiguity of post psychotic depression. Schizophr Bull 4: 477-480, 1978.
6. Pope HG, Lipinski JF: Diagnosis in schizophrenia and manic depressive illness. Arch Gen Psychiatry 35: 811-828, 1978.
7. Bland RC, Orn N: Schizophrenia: Diagnostic criteria and outcome. Br J Psychiatry 134: 34-38, 1979.
8. Gift RE, Strauss JS, Kokes RF et al: Schizophrenia: Affect and outcome. Am J Psychiatary 137: 580-585, 1980.
9. Johnson DAW: Studies of depressive symptoms in schizophrenia. I. The prevalence of depression and its possible causes. Br J Psychiatry 139: 89-101, 1981.
10. Falloon I, Watt DC, Shepherd M: A comparative trial of pimozide and fluphenazine deconoate in continuation therapy of schizophrenia. Psychol Med 8: 59-70, 1978.
11. Miles P: Conditions predisposing to suicide: a review. J Nerv Ment Dis 164: 231-246, 1977.

12. Tsuang MT: Suicide in schizophrenics, manics, depressives and surgical controls. Arch Gen Psychiatry 35: 153-155, 1978.
13. Black DW, Winokur G, Warrack G: Suicide in schizophrenia: The Iowa Record Linkage Study. J Clin Psychiatry 46: 11 (sec 2) 14-17, 1985.
14. Siris SG, Rifkin A, Reardon GT et al: Stability of the post psychotic depression syndrome. J Clin Psychiatry 47: 86-88, 1986.
15. Siris SG, Rifkin A, Reardon GT et al: Course related depressive syndromes in schizophrenia. Am J Psychiatry 141: 1254-1257, 1984.
16. Brown SJ, Sweeney DR, Schwartz GE: Differences in self reported and observed pleasure in depression and schizophrenia. J Nerv Ment Dis 167: 410-415, 1979.
17. Craig TJ, van Natta P: Recognition of depressed affect in hospitalized psychiatric patients: staff and patient perceptions. Dis Nerv Sys 37: 561-566, 1976.
18. Strian F, Heger R, Klicperac: The course of depression for different types of schizophrenia. Psychiatric Clinics (BASEL) 14: 205-214, 1981.
19. van Putten T, May PR: Akinetic depression in schizophrenia. Arch Gen Psychiatry 35: 1101-1107, 1978.
20. Rifkin A, Quitkin F, Klein DF: Akinesia, a poorly recognized drug induced extrapyramidal behavioural disorder. Arch Gen Psychiatry 32: 672-676, 1975.
21. Prosser ES, Csernansky JG, Kaplan J et al: Depression, parkinsonian symptoms and negative symptoms in schizophrenics treated with neuroleptics. J Nerv Ment Dis 175: 100-105, 1987.
22. McGlashan T, Carpenter WT: An investigation of the post psychotic depressive syndrome. Am J Psychiatry 133: 16-19, 1976.
23. Knights A: "Revealed" depression and drug treatment for schizophrenia. Arch Gen Psychiatry 38: 806-811, 1981.
24. Ndetei DM, Singh A: Schizophrenia with depression: Causal or Coexistent. Br J Psychiatry 141: 354-356.
25. Prusoff BA, Williams DH, Weissman MM et al: Treatment of secondary depression in schizophrenia. Arch Gen Psychiatry 36: 569-575, 1979.
26. Mandel MR, Severe JB, Schooler NR et al: Development and prediction of post psychotic depression in neuroleptic treated schizophrenics. Arch Gen Psychiatry 39: 197-203, 1982.
27. Lerner Y, Moscovich D: Depressive symptoms in acute schizophrenic hospitalized patients. J Clin Psychiatry 46: 483-486, 1985.
28. Munro JG, Hardiker TM, Leonard DP: The dexamethasone suppression test in residual schizophrenics with depression. Am J Psychiatry 141: 250-252, 1984.
29. Roy A, Thompson R, Kennedy S: Depression in chronic schizophrenics. Br J Psychiatry 142: 465-470, 1983.
30. Siris S, Rifkin A, Reardon G et al: The dexamethasone suppression test in patients with post psychotic depression. Biol Psychiatry 19: 1351-1356, 1984.
31. Siris SG, Morgan V, Fagerstrom R et L: Adjunctive imipramine in the treatment of post psychotic depression. Arch Gen Psychiatry 44: 533-539, 1987.
32. Craig TJ, Richardson MA, Pass R et al: Measurement of mood and affect in schizophrenic inpatients. Am J Psychiatry 142: 1272-1277, 1985.

33. Wing JK, Cooper JE, Sartorius N: <u>The Measurement and Classification of Psychiatric Symptoms</u>. London: Cambridge University Press, 1974.

34. Carroll BJ, Feinberg M, Greden J et al: A specific laboratory test for the diagnosis of melancholia, <u>Arch Gen Psychiatry</u> 38: 15-22, 1981.

35. Andreasen NC, Olson S: Negative versus positive symptoms in schizophrenia: Definition and validation. <u>Arch Gen Psychiatry</u> 309: 789-794, 1982.

36. Hamilton M: A rating scale for depression. <u>J Neurol Neurosurg Psychiatry</u> 23: 56-62, 1960.

37. Carpenter WT, Heinrichs DW, Wagman AM: Deficit and nondeficit forms of schizophrenia: The Concept. <u>Am J Psychiatry</u> 145: 578-583, 1988.

DEPRESSION IN FUNCTIONAL PSYCHOSIS

D.G.C. Owens and Eve C. Johnstone

Northwick Park Hospital and
Clinical Research Centre, Harrow, Middlesex,
England

Despite a recent reappraisal of the concept of Unitary Psychosis (1), the Kraepelinian dichotomy of functional psychotic illnesses remains the theoretical and clinical orientation of most psychiatrists. The validity of this in defining separate disease entities is, of course, crucially based on outcome. Thus in establishing a diagnosis, Kraepelin urged consideration of...

> "the entire picture ... especially ... the
> changes which it undergoes in the course of
> the disease" (2).

Such longitudinal methodology has, however, largely been abandoned in favour of the cross-sectional evaluation of symptomatology. Schneider, for example, stated that:

> "Psychiatric diagnosis must be based on the
> presenting situation, not on the course
> taken by the illness" (3).

A major problem with cross-sectionally based diagnosis of functional disorder arises with the evaluation of affective symptoms, particularly depression. While both Kraepelin and Bleuler recognized the prominence of depression in schizophrenia -- Bleuler called it "one of the most frequent acute disturbances in schizophrenia" (4) -- progressive modifications of the schizophrenia concept, and in particular the widespread interest in the schizoaffective category, have resulted in the gradual exclusion of depression as a core symptom of schizophrenia (5). Recent studies have indicated that this is an important omission, as depression in this context appears to be common (c.f. Table 4) and may have greater implications for both morbidity (6,7,8) and mortality (9,10) than has hitherto been appreciated.

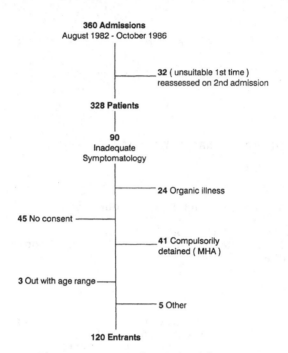

360 Admissions
August 1982 - October 1986

32 (unsuitable 1st time)
reassessed on 2nd admission

328 Patients

90
Inadequate
Symptomatology

24 Organic illness

45 No consent

41 Compulsorily
detained (MHA)

3 Out with age range

5 Other

120 Entrants

Figure 1. Sample Selection

The present work is concerned with both depression of mood and other depressive features occurring in a group of patients all of whom suffered from functional psychosis. The sample comprised the 120 patients who participated in a placebo controlled in-patient study of the efficacy of neuroleptic and/or lithium in the treatment of functional psychotic illness (11). While subjects were clearly selected on the basis of their participation in the drug trial, the study has the advantage that the patients were not specifically diagnosed at the time of their initial assessments. Standardized diagnostic criteria (DSM III) were applied retrospectively at the end of the trial. Hence, the initial ratings were not biased by diagnostic assumptions. Data will be presented on the prevalence and severity of depressive abnormality both overall and within diagnostic groups and some limited results will also be presented on the status of the sample at follow-up approximately 2 years later.

METHODS

Patient Selection

From August 1982 all consecutive admissions to the Clinical Research Centre, Division of Psychiatry who had definite or possible psychotic illness were considered for inclusion in the trial. On admission, in addition to a conventional historical enquiry and mental state assessment, all potential entrants had a Present State Examination (PSE)

(12) performed and were considered eligible for inclusion in the drug study if they scored unequivocal abnormality on at least one of the psychotic items of the PSE (including those reflecting thought disorder) and had no evidence of organicity underlying their mental state abnormality. Patients from whom consent had been obtained were entered up to a total of 120 participants. These trial patients -- all of whom suffered from functional psychotic illness -- are the subjects of the present study.

Non-Participants

The study group represent 56.9% of those _initially conforming to inclusion criteria_. Those not included comprised 45 cases who refused consent, 41 cases compulsorily detained under the Mental Health Act and 5 cases excluded for a variety of reasons including being mute or having an inadequate grasp of English (Figure 1).

Ratings and Diagnostic Classifications

On the day prior to commencement of treatment (and twice weekly during the trial) subjects were rated in terms of their mental status and extrapyramidal neurological signs. The schedules that were completed and are relevant to the present work were:

1)   Krawiecka Scale (K) (13).
2)   Montgomery-Asberg Scale (M-A) (14) - for depression.
3)   Abnormal Involuntary Movement Scale (AIMS) (15) - for hyperkinetic disorder.
4)   Scale for Targeting Abnormal Kinetic Effects (TAKE) (16) for features of Parkinsonism.

The Krawiecka Scale is a simple and reliable method of rating mental state in psychotic patients. Although the present work relates mainly to depression, reference is made to "positive" and "negative" mental state features. These are the sum of Krawiecka Scale ratings for hallucinations/ delusions/ incoherence/ affective incongruity and affective flattening/ poverty of speech respectively. The Montgomery-Asberg Scale is particularly sensitive to measuring change in patients in therapeutic trials and it is not a requirement that it be applied to pre-diagnosed cases. In addition to rating objective (apparent) and subjective (reported) sadness, it rates 8 features of depressive illness -- inner tension/reduced sleep/reduced appetite/concentration difficulties/lassitude/inability to feel/ pessimistic thoughts/suicidal thoughts. Both these scales are ordinal and, in addition to rating the presence of abnormality, allow an assessment of its degree (severity continuum for Krawiecka 0-4 and for Montgomery-Asberg 0-6).

After completion of the drug study and at a point before follow-up, DSM III criteria (17) were applied to the casenotes of the sample and diagnoses valid at the point of entry were established.

After approximately 2 years the sample was traced and where possible re-examined in terms of the same schedules used at initial contact.

RESULTS

Table I shows the distribution of diagnoses along with the age and sex distribution in each group, after retrospective application of DSM III criteria. In all there were 64 males and 56 females, the mean age of the total sample being 35.1 years. Manic patients were somewhat younger and those with paranoid disorder (although few in number) relatively older than those in other diagnostic categories. There were twice as many male schizophrenics as female. In view of the small number fulfilling criteria for atypical psychosis (7) and paranoid psychosis (5) these categories have been combined throughout (A/P).

Initial Contact

At initial contact significantly greater degrees of depressed mood were found in those subsequently fulfilling DSM III criteria for major depressive episode (KRAWIECKA SCALE: H = 39.7, p < 0.001 [KRUSKALL-WALLIS]: Post hoc pairwise comparison [SHEFFE'S METHOD] D > S: p < 0.05. D > SP: p < 0.001. D > SA: p < 0.01/ D > M: p < 0.001. MONTGOMERY-ASBERG SCALE: H = 35.5: p < 0.001 [KRUSKALL-WALLIS] Post hoc pairwise comparison [SHEFFE'S METHOD] D > S: p < 0.05. D > SP: p < 0.01. D > SA: p < 0.01. D > M: p < 0.001). 'Positive' mental state features -- present to some extent in all by the terms of their inclusion -- were found to greater degree in the schizophrenics (H = 11.5, p < 0.025), as were 'negative' features (H = 13.8: p < 0.01 [KRUSKALL-WALLIS]) the major difference here lying in the much lesser abnormality found in the manics (S > M: p < 0.01 [SHEFFE'S METHOD]). Neither AIMS or TAKE scores related to DSM III classifications. There was no difference in the rating of depression of mood between the sexes.

Using data from the Krawiecka Scale ratings, clinically significant depression of mood (K ≥ 2) was present in 42.5% of the total sample and was found across all DSM III categories, with the exception of mania where it was minimally represented (Figure 2). Of the DSM III schizophrenics, 35% were depressed by this criterion. Comparison of the severity distributions (Figure 3) shows that while depression was usually of moderate or severe degree in the depressives, it was by contrast only mild or moderate in severity in the schizophrenics, none of whom were rated as severe at presentation. The mean K depression scores in each diagnostic category are shown in Figure 4 (Mean scores are illustrated with standard errors throughout).

Those in 'schizophrenic' groups (DSM III schizophrenia, schizophreniform, atypical and paranoid psychoses) did not significantly differ from each other nor from those fulfilling schizoaffective criteria. In terms of severity

Table 1. The Total Sample (N = 120)

DSM III	N	MEAN AGE (SD)	N AND MEAN AGE (SD) BY SEX		
ˣSchizophrenia (s)	49	35 (10.7)	M (33)	32.9	(9.6)
			F (16)	39.2	(11.9)
ˣSchizophreniform Psychosis (SP)	15	31 (9.6)	M (5)	26.8	(6.1)
			F (10)	33.1	(10.6)
ˣParanoid Disorder (PD)	5	49 (6.4)	M (1)		
			F (4)		
ˣAtypical Psychosis (AP)	7	38.9 (19.2)	M (5)		
			F (2)		
Schizoaffective Psychosis (SA)	10	38.1 (10.8)	M (5)	32.4	(4.3)
			F (5)	43.8	(12.7)
Mania (M)	18	23.1 (11.7)	M (8)	30.4	(10.1)
			F (10)	33.5	(13.2)
Depression (D)	16	34.9 (13.2)	M (7)	33.3	(12.7)
			F (9)	36.1	(14.1)
All 'Schizophrenic'	76	35.4 (11.8)	M (44)	31.9	(10.1)
			F (32)	40	(12.6

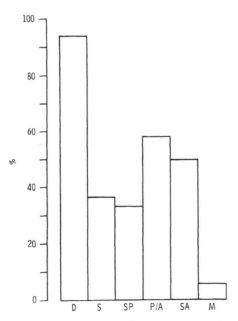

Figure 2. Prevalence of Depression (KR 2 or more) - Initial Contact

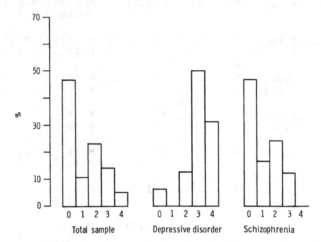

Figure 3. Distribution of Krawiecka Depression Scores – Initial Contact

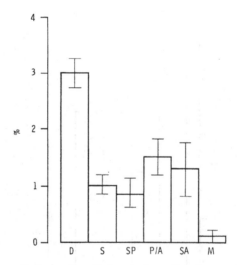

Figure 4. Mean Krawiecka Depression Scores – Initial Contact

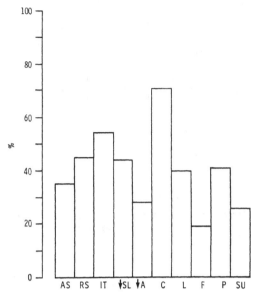

Figure 5.   Frequency of Positive Rating (2 or more) on Items of Montgomery-Asberg Scale - Initial Contact

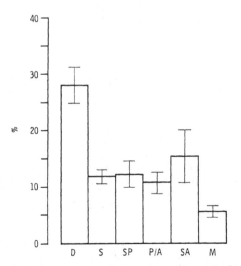

Figure 6.   Mean Montgomery-Asberg Scores - Initial Contact

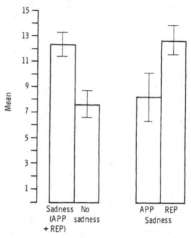

Figure 7. Mean of M-A Items 3-10 in Relation to Sadness
(Items 1-2) in Non-Affective Psychotics

of depression such schizoaffective patients by DSM III
criteria are as a group more like schizophrenics than those
with major depressive episodes.

The distribution of a positive rating (> 2) on each M-A
item is shown in Figure 5. 'Impaired concentration' was the
most frequently rated abnormality (70.8%). In half of these
(34.2% of total) the disorder was marked. 'Inner tension'
was found in 54.2% of patients. While half the
schizophrenics rated on this, the great majority of the
depressives (88.2%) scored 2 or more, though only 22.2% of
manics scored comparably.

'Reported sadness' was present in 45.9% with 7
schizophrenics rating 4 (out of a maximum of 6) but none
higher. By contrast, 13 depressives scored 4 or more.
Thirty five percent had 'apparent sadness'. Although 9
schizophrenics (out of 49) scored 4 on 'pessimism', severe
degrees of this were almost exclusively confined to depressed
subjects, 9 of whom (56.3%) rated 4 or more. Likewise
'suicidal ideation' - present in 25.8% of the total sample --
was in its severe form largely a feature of depression ($\geq$ 4
in 25% D and 14.6% S). It is, however, worth noting that
although none of the schizophrenics had severe depression of
mood, 3 nonetheless had extremely prominent suicidal
ideation.

The least commonly rated M-A item was 'inability to
feel' (20.8%). While 37.5% of depressed patients rated 2 or
more, this was only noted in 18.4% of schizophrenics. In no
patient was 'inability to feel' present in severe degree.

As with K depression ratings, the mean M-A scores did not significantly differ in those in 'schizophrenic' categories (DSM III S, SP, A/P) nor between them and those with schizoaffective psychosis, while all these groups differed significantly from affective diagnoses (Figure 6).

Clearly with a multi-item scale such as the M-A, it is possible to obtain a positive rating in the absence of either apparent or reported sadness. Figure 7, which shows the means for M-A items 3-10 in all the non-affective patients with and without depression, confirms this. However, the mean for these items obtained in those with mood change (apparent and reported sadness) is significantly greater than that of the patients with no mood change (t = 3.2, df = 65, p < 0.01). Furthermore, Figure 7 also illustrates that this finding is due to the increase in mean scores of those who report subjective sadness. The mean of items 3-10 for those rated as appearing sad but not as feeling sad is not significantly different from that in those with neither sadness item rated.

In view of the high prevalence of depression in the schizophrenic patients a comparison was made of depressive features (M-A items 3-10) in schizophrenics without depressed mood, schizophrenics with depression and those patients with major depressive episodes (Table 2). In view of the above findings 'depression' was defined as reported sadness of 2 or more. While there was a tendency for mean item scores to increase progressively through these respective categories, few of the differences were significant. Thus, in comparing non-depressed with depressed schizophrenics, only 'pessimistic thoughts' and 'suicidal ideation' were significantly greater in the latter group, while comparison of the depressed schizophrenics with major depressives showed only 'inner tension' and 'pessimism' to be greater in the latter group (though the increase in 'sleep disturbance' in the major depressives narrowly missed significance). 'Impaired concentration' and 'inability to feel' were not significantly less in evidence in the non-depressed schizophrenics than the major depressed patients, but all other items were highly significantly so (1% - 0.01% level).

In recent years attention has been drawn to the apparent similarity between a number of the clinical deficits to be found in the sorts of patients comprising a large proportion of the present sample (18,19). Specifically, the potential difficulty in separating for rating depression, flattened affect and neuroleptic-related Parkinsonism has been emphasized. In the present sample both 'negative' abnormality and Parkinsonian features were not infrequently encountered and the question arises as to whether ratings of 'depression' could have been contaminated by these clinically similar phenomena. Table 3 shows the significant correlations obtained between relevant ratings from various schedules used. No correlations emerged between Krawiecka ratings of depression or Montgomery-Asberg scores and either 'negative' mental state abnormality or features of parkinsonism as rated on the TAKE, which would tend to imply that these features were in general rated independently.

Table 2. Mean (+ SEM) M-A Scores for Items 3-10 in Schizophrenics with and Without Depression and Patients with Major Depressions

	Schizophrenics Without Depression	Significance	Schizophrenics With Depression	Significance	Major Depressive Episode
Inner Tension	1.03 ± 0.27	$p < 0.06 > 0.05$	1.9 ± 0.32	$p < 0.05$	2.93 ± 0.38
Reduced Sleep	0.55 ± 0.27	N/S	1.25 ± 0.38	$p < 0.01 > 0.05$	2.44 ± 0.47
Reduced Appetite	0.41 ± 0.21	N/S	0.95 ± 0.29	N/S	1.88 ± 0.51
Concentration Difficulties	2.45 ± 0.30	$p < 0.1 > 0.05$	3.25 ± 0.23	N/S	3.19 ± 0.41
Lassitude	0.86 ± 0.27	$p < 0.1 > 0.05$	1.65 ± 0.32	N/S	2.56 ± 0.43
Inability to Feel	0.48 ± 0.24	N/S	0.8 ± 0.33	N/S	1.31 ± 0.45
Pessimistic Thoughts	0.41 ± 0.19	$p < 0.0001$	2.25 ± 0.41	$p < 0.02$	3.88 ± 0.46
Suicidal Thoughts	0.38 ± 0.22	$p < 0.001$	2.05 ± 0.46	N/S	2.06 ± 0.42

Table 3. Correlations Between Examination Variables

	DEP	ANX	RET	POS	NEG	BR*	MA	AIMS	TAKE
Krawiecka Depression									
Krawiecka Anxiety	+ $p<0.0001$								
Krawiecka Retardation	+ $p<0.001$	+ $p<0.01$							
Krawiecka 'Positive'									
Krawiecka 'Negative'		+ $p<0.05$	+ $p<0.0001$						
Bech* Rafaelsen	- $p<0.001$	- $p<0.01$	- $p<0.01$		- $p<0.01$				
Montgomery Asberg	+ $p<0.0001$	+ $p<0.0001$	+ $p<0.0001$			- $p<0.01$			
AIMS									
TAKE								+ $p<0.0001$	

* Bech P, Rafaelsen OJ, Kramp P et al: The mania rating scale: scale construction and inter-rater reliability. Neuropharmacology, 17: 430-431, 1978.

Table 4. Depression in Schizophrenia Literature

Author	Patient Type	N	Criteria for Schizophrenia	% with Depression	Comment
Helmchen & Hippius (20) (1976)	In course of neuroleptic treatment	120		50%	
McGlashan & Carpenter (23) (1976a)	Various	174	Various	25%	Estimate from review of 4 studies
McGlashan & Carpenter (24) (1976b)	Discharged	30	PSE/DSM III	43.3% 50%	At discharge at 1 year
Weissman et al (25) (1977)	Out-patients	50	New Haven DSM II	28%	
Planansky & Johnston (33) (1978)	Chart review (in-patient)	115	?	56.5%	Wide concept of depression (incl. guilt, worthlessness)
Cheadle et al (26) (1978)	Mainly in remission 'small no. in hospital'	157	'Current British criteria'	36.3%	For PSE 'simple depression'
Falloon et al (6) (1978)	Acute in-patients and at 1 year	44	PSE	40.9%	Same for in-patient and during trial
Knights et al (27) (1979)	Acutes followed over 6 months	37	PSE	54%	High drop out rate

Study	Setting	N	Diagnostic criteria	%	Criteria for depression
Roy (34) (1980)	Chart review (in-patients)	100	DSM III chronic paranoid schizophrenia	30%	DSM III criteria for major depression (2u)
Siris et al (28) (1981)	Chart review (in-patients but post-)	50	RDC	40% 28%	'Sadness' RDC criteria
Johnson (35) (1981)	First episodes Chronics (in relapse)	67 168	Schneider "	26.9% 34.5%	
Strian et al (36) (1982)	In-patients	134	ICD 9	59.7%	Self ratings
Mandel et al (29) (1982)	Out-patients followed for max. of 1 year	211	DSM III	22.3	
Guze et al (30) (1983)	? Out-patients	44	Feighner-esque	56.8%	Long-term follow-up (6-12 years)
Curzon et al (31) (1985)	Chronic out-patients	62	PSE	24.4%	PSE 'simple depression'
Drake et al (40) (1986)	In-patients	104	DSM III	54% 21%	'Depressed mood' DSM III major depressive episode
Elk et al (41) (1986)	In-patients	56	PSE	30%	
Berrios & Bulbena (5) (1987)	Chart review	78	ICD	37.2%	'Post psychotic depression'

Table 4. Continued

Follow-up

A total of 105 patients (87.5% of the initial sample) were examined at follow-up. The average follow-up interval was 27.5 months.

Two deaths had occurred at the point of follow-up and a further two by the end of the follow-up interval. One patient with paranoid disorder died of malignant melanoma and a further 3 (1D, 1S, 1M) killed themselves.

At follow-up neither the presence of depression (K) nor of depressive features (M-A) significantly related to depression or DSM III diagnoses at initial contact. While there was a tendency for 'positive' features at follow-up to be more common in those initially diagnosed as schizophrenic and schizoaffective and less so in schizophreniform and manics, and for AIMS scores to be higher in schizophrenics, none of these differences attained conventional levels of significance. Likewise, neither 'negative' scores nor TAKE scores related to the previously applied DSM III classifications. In terms of the M-A scale, however, women had significantly lower scores than men (p < 0.05).

At follow-up depression had eased substantially in all groups, with a K ≥ 2 being noted in only 17.1% of the sample. Of the major depressive episode patients 30% still showed depressed mood at follow-up but this was never severe, the comparable figure for schizophrenics being 15.3% (Fig. 8).

The mean Krawiecka depression scores fell substantially in all diagnostic categories (Fig. 9) though this decline was only significant for the major depression patients (t = 5.11, df 24, p < 0.0001), the schizophrenics (t = 2.07, df 93, p < 0.05) and the paranoid/atypical psychotics combined (t = 2.48, df 21, p < 0.05). The manics had, of course, barely detectable degrees of depression initially and the failure of the schizoaffectives to fall significantly is largely due to the influence of a single patient rated mild at first contact but whose depression was severe at follow-up. The reduction in depression was greatest in the major depressives (77%) but only (43%) in the schizophrenics.

All the individual M-A items except 'inability to feel' showed substantial reductions and in particular severe degrees of 'pessimism' and 'suicidal ideation' were not found at follow-up. As Fig. 10 shows, all diagnostic groups ended up with essentially comparable mean M-A scores. Except for the manics and schizoaffectives, these represent significant reductions in all groups (D: t = 5.26, df 24, p < 0.0001. S: t = 2.33, df 93, p < 0.025. SP: t = 2.81, df 27, p < 0.01, P/A: t + 2.82, df 21, p < 0.025). The greatest falls were again in the depressed patients (82.1%), but only 47.5% in schizophrenics.

Thus, whatever had been or was being done for these patients had resulted in a substantial improvement in at least these aspects of the mental state. This, however, does not mean they were otherwise entirely well (Fig. 11). Thirty four percent of the follow-up sample remained symptomatic in

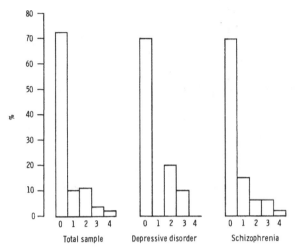

Figure 8.   Distribution of Krawiecka Depression Scores -
            Follow-up

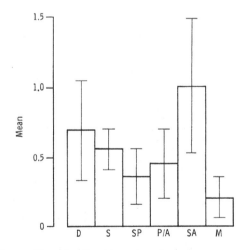

Figure 9.   Mean Krawiecka Depression Scores - Follow-Up

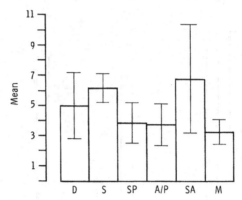

Figure 10.   Mean Montgomery-Asberg Scores - Follow-Up

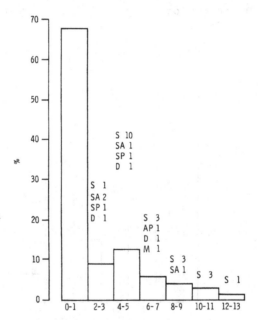

Figure 11.   Distribution of Krawiecka 'Positive' Scores -
Follow-Up

terms of 'positive' mental state features. As might be expected the great majority of these (65.6%) were schizophrenics.

DISCUSSION

The major limitation of this study has already been noted in that patients were selected on the basis of their participation in a therapeutic trial. While this may restrict the degree to which the findings can be generalized the sample was nonetheless demographically fairly typical of those admitted to hospital with functional psychotic illness. The potential problem of selection notwithstanding, the population demonstrated high prevalences of both depressed mood, and features of depression defined in terms of the scale used. Almost half this sample had clinically significant depression at presentation as had, most notably, just over one third of those subsequently classified as schizophrenic. Correspondingly high prevalences of depressive features were also found. This amounts to a substantial body of psychopathology easy to overlook in the setting of florid psychotic abnormality.

The major interest in the present work lies in the mental state comparisons possible across diagnoses, the psychopathology having been rated in standardized fashion at a time prior to the application of diagnostic criteria. Hence, the ratings were comprehensive in all cases and to a large extent free from diagnostic bias.

After many years of neglect, there has been in the past decade increasing interest in depression occurring in the context of schizophrenia (Table 4). This was sparked by concern in the American literature about so-called post-psychotic depression (5,6,20-31) though it is clear that depressed mood is not confined to the post-psychotic phase of symptom resolution but can also be found in the prodromal or pre-relapse phase (32) as well as the acute, florid phase (6,33-41). Whether depressions associated with different phases of illness have shared or divergent aetiologies is unknown.

Studies of the prevalence of depression in schizophrenia (Table IV) are difficult to compare. While most of these studies adopt strict and standardized criteria for schizophrenia, what is meant by depression is often unstated or widely variable. Furthermore, some studies report point prevalences while others are concerned with period prevalence. Approximately one half were concerned with depression in the remission phase only (post-psychotic depression). Nonetheless, it does appear that between 25-40% of schizophrenic patient experience depression during the in-patient (acute) phase of illness, the higher figure applying to depressed mood per se, the lower to depressive syndrome more rigorously defined. Longer follow-up periods or novel techniques such as self-ratings, may enhance these figures. This is in keeping with the (point) prevalence of 35% reported here.

It has been argued that a number of the features used to diagnose depression in schizophrenics are actually features of schizophrenia itself, and that depression in schizophrenics differs qualitatively from that found in patients with primary major depressive illness (42). While it may be argued that in terms of the Montgomery-Asberg Scale 'impaired concentration' and 'inability to feel' are of little value in distinguishing between schizophrenia and depression, the present results do not point clearly to such a qualitative difference. All items on the M-A scale were positively represented in the schizophrenics (both depressed and non-depressed) as in the depressives. The major differences seem to be largely ones of degree. The impression emerged of a progression in mean M-A item scores from schizophrenics without depression, through depressed schizophrenics to patients with major depressive episodes (Table 2).

The view that depression in schizophrenia, while common, is qualitatively similar to primary depression is supported by the findings of Weissman et al (25), who compared the symptomatology of patients with primary depression, with that occurring as a secondary phenomenon in other disorders including schizophrenia. From objective and self-report ratings, they concluded that both groups had relatively similar symptom patterns which differences in severity. Reported depression was less severe in the secondary depressives who were rated less symptomatic by clinicians. The objective assessments were particularly noted as milder in the schizophrenics. The fact that in the present study those classified as having 'major depressive episode' had also, by virtue of fulfilling trial criteria, additional psychotic features in their mental state, does not alter the same basic conclusion being drawn.

'Inability to feel' was one of the least useful items in separating depressed patients from either schizophrenics with or those without depressed mood. This may relate to some extent to the problem of separating depression from the anhedonia frequently found particularly in established schizophrenia - a distinction that may be equally difficult for the patient as for the examiner. Likewise 'impairment of concentration' did not significantly differ between the non-depressed schizophrenics and the major depressives reflecting the relative non-specificity of this abnormality in major mental state disorder. The most important single abnormality associated with depressed mood was 'pessimism' which showed a progressive and highly significant increase in degree from non-depressed through depressed schizophrenics to patients with major depression. Suicidal ideation was equally prominent in the depressed schizophrenics as in the major depressive group. This is of particular note since it must be recalled that none of the schizophrenics were rated as having severely depressed mood. The finding is accounted for by a small number of schizophrenics with severe and prominent suicidal ideation in the presence of only mild to moderate depression. The reasons for this are unclear and may relate to the influence of psychotic phenomena (delusions or imperative hallucinations) in the relative absence of mood change, or to an erosion of affect as a result of the schizophrenic process. Such patients clearly represent a high risk group easily overlooked in the routine clinical setting.

In enhancing the severity of the combined M-A ratings (i.e. items 3-10) in the schizophrenics the important factor is the degree of depression reported by the patient rather than the apparent depression appraised by the examiner. It remains unclear whether this is a finding unique to schizophrenia, where the expression of mood disorder may be aberrant or concealed, or whether it is more generalizable to other secondary depressive states.

In general, the other non-affective psychotic categories (schizophreniform psychosis, atypical psychosis and paranoid disorder) did not differ from the schizophrenic patients in the degree of their depression (mean K) or depressive features (mean M-A). This is also true of the schizoaffective subjects, who in this regard were clinically closer to the schizophrenic than affective patients. The number of schizoaffective patients on the present sample was small and whether this is a genuine finding that can be generalized requires confirmation. Furthermore, it may be that schizoaffective patients show different degrees of affective features at different phases of illness and the above finding may relate to the fact that subjects were rated within 2 weeks of admission when psychotic features may have dominated. Although the present sample were retrospectively classified in terms of the original DSM III criteria, the revised version of these (43) may make such a finding more likely in future as they emphasize the temporal association between schizophrenic and mood symptoms. In any given episode, psychotic symptoms must be present for at least 2 weeks without 'prominent' mood disorder for the criteria to be fulfilled and hence patients assessed early in their episode would be expected to produce relatively lower mood ratings.

The association of Krawiecka depression ratings and Montgomery-Asberg scores with a subsequent DSM III diagnosis of major depressive episode was to be expected, as was the association between 'positive' and 'negative' mental state features and schizophrenia. It is perhaps surprising, however, that no relationship emerged between AIMS and TAKE scores and the category (or categories) most likely to be associated with antipsychotic drug use -- i.e. schizophrenia. It may be that these preparations are more widely prescribed to patients of the sort described here or that their extrapyramidal side-effects are more durable than is appreciated or it may be that the ratings were contaminated by non-drug related phenomena. This raises the crucial question of the ability of examiners to separate clinically several of the types of deficit exhibited in such patients as these -- in particular depression, 'negative' mental state features and Parkinsonism which have been reported to overlap (44). While in statistical terms in the present sample there was no significant cross-contamination (Table 3), this does not rule out the possibility of an overlap either in a small number of individual patients or in general with disorders of mild degree.

One of the major unresolved dilemmas in relation to depression in schizophrenia is the possible role of anti-psychotic medication in promoting or causing the depressed state (45-49). Much of the argument in favour of a causal connection between the two has focussed on post-psychotic depression but is based largely on anecdotal or uncontrolled observations (45), and studies which show a movement towards resolution of depression in schizophrenics during the course of treatment of an acute phase of illness (33,36,37,47) argue strongly for there being no causal association. The present study cannot directly address this question. It can be said, however, that using arbitrary exposure criteria ('off' neuroleptics for at least 2 months continuously prior to assessment/'on' for at least 2 weeks continuously prior to assessment), neither mean Krawiecka depression scores nor mean Montgomery-Asberg scores were significantly higher in those 'on' neuroleptics at assessment in i) the total sample, ii) the DSM III schizophrenics alone or iii) all 'schizophrenic' categories combined (S+SP+P+A). Likewise those 'on' and 'off' neuroleptics did not differ significantly in their mean 'positive' feature scores, 'negative' feature scores or overall anxiety. They did, however, differ in their degree of extrapyramidal disorder. Thus, those receiving neuroleptics had significantly higher mean AIMS (t = 2.07, df 47, p < 0.05) and TAKE (t = 3.52, df 47, p < 0.001) scores.

The concept of akinetic depression has been widely debated since its introduction (50) but its heuristic value remains uncertain. The present findings do not support its validity but again suggest that by careful clinical assessment the mental state deficits associated with the schizophrenic illness can in general be separated from the features of Parkinsonism that relate to its treatment.

The suicide rate is rather low in this group though the follow-up interval is relatively brief. The topic of suicide in such patients is dealt with in detail elsewhere in this volume and the only point that will be noted here is that two of the three patients who killed themselves were graduates, emphasizing the risk in such individuals, and although one was manic at index admission, he killed himself (with the novel method of chewing the leaves of the yew tree) while depressed.

Follow-up data must be treated with caution as the role of treatment since index admission could not be taken into account. This is largely because on discharge a number of subjects entered an on-going out-patient phase of the drug study and hence were maintained on trial medication at point of follow-up, the examiners remaining blind to the nature of this. Nonetheless, the results do provide an interesting 'snap-shot' of the patients' mental states after approximately 2 years, especially in terms of the degree and pattern of change.

Overall, depression had resolved substantially at follow-up. This was particularly true of those diagnosed as having major depressive episodes initially. Their mean depression ratings (Krawiecka) declined by 77% while their mean Montgomery-Asberg scores fell by 82.1%. Both these represent very significant decreases in psychopathology. In fact it is possible that improvement in the depressed group was even greater since this group suffered the highest attrition rate to follow-up, with only 62.5% available for interview. If this could be explained by a higher percentage of these patients being well and unwilling to maintain psychiatric contact the improvement for the group as a whole would appear even more substantial.

It is interesting to note that while the schizophrenics also significantly reduced both these mean scores (K and M-A) the degree of their reductions was not so great. The impression was of a much less predictable resolution in depression in the schizophrenics, perhaps from a less complete or less rapid reduction in this type of symptomatology. This is in keeping with the findings of Shanfield et al (22) who noted that depression resolved more slowly in their schizophrenic than their depressed patients and is compatible with differential rates of resolution of depression in schizophrenics reported by McGlashan and Carpenter (24). It must, however, be emphasized that the finding here refers to a group effect. Depression at follow-up could not overall be predicted by depression at initial contact and it may be that there exists within the schizophrenic group a greater fluidity of depressive symptomatology -- with features resolving in some as they develop in others. In those with major depression, on the other hand, the tendency may be toward a more predictable resolution, at least over the time span of the present follow-up.

The present study demonstrates a high prevalence of depression in functional psychotics. This is pertinent to schizophrenia, the diagnosis of which can clearly be sustained even in the presence of depression of mood. While depression itself may not usually be severe, it may be associated with a number of features, severe in themselves, which may have major implications for the patients.

It is easy in psychiatry to think 'on tram lines' particularly with psychotic patients who may have an impaired ability to express the subjective roots of their distress or may be inconsistent or unreliable witnesses. A relatively few clinical features may be sufficient to establish a diagnosis even using standardized operational criteria, but to restrict ones efforts merely to eliciting sufficient information to establish a diagnosis is to risk ignoring considerable areas of morbidity in the patient's mental life.

REFERENCES

1. Crow TJ: The continuum of psychosis and its implication for the structure of the gene. Br J Psychiatry 149: 419-429, 1986.
2. Kraepelin E: Dementia Praecox. Trans. R.M. Barclay. Robert E. Krieger. Huntington, New York, p. 257, 1971.
3. Schneider K: Clinical Psychopathology. Trans. M.W. Hamilton, Grune and Stratton, New York, London, p. 92, 1959.
4. Bleuler E: Dementia Praecox or the Group of Schizophrenias. Trans. Joseph Zinkin. International Universities Press, New York, p. 208, 1950.

5. Berrios GE, Bulbena A: Post psychotic depression: The Fulbourn cohort. Acta Psychiat Scand 76: 89-93, 1987.
6. Falloon I, Watt DC, Shepherd M: A comparative controlled trial of pimozide and fluphenazine decanoate in the continuation therapy of schizophrenia. Psychol Med 8: 59-70, 1978.
7. Johnson DAW: Studies of depressive symptoms in schizophrenia II: A two-year longitudinal study of symptoms. Br J Psychiatry 139: 93-96, 1981.
8. Glazer W, Prusoff B, John K et al: Depression and social adjustment among chronic schizophrenic out-patients. J Nerv Ment Dis 169: 712-717, 1981.
9. Roy A: Suicide in chronic schizophrenia. Br J Psychiatry 141: 171-177, 1982.
10. Drake RE, Gates C, Cotton PG et al: Suicide among schizophrenics: Who is at risk? J Nerv Ment Dis 172: 613-617, 1984.
11. Johnstone EC, Crow TJ, Frith CD et al: The Northwick Park 'functional' psychosis study: Diagnosis and treatment response. Lancet ii: 119-125, 1988.
12. Wing JK, Cooper JE, Sartorius N: The Measurement and Classification of Psychiatric Symptoms. Cambridge University Press. London, 1974.
13. Krawiecka M, Goldberg D, Vaughan M: A standardised psychiatric assessment scale for rating chronic psychotic patients. Acta Psychiat Scand 55: 299-308, 1977.
14. Montgomery SA, Asberg M: A new depression scale designed to be sensitive to change. Br J Psychiatry 134: 382-389, 1979.
15. Guy W: ECDEU Assessment Manual for Psychopharmacology. DHEW Washington DC, p. 534-537, 1976.
16. Wojcik JD, Gelenberg AJ, Labrie RA et al: Prevalence of tardive dyskinesia in an out-patient population. Compr Psychiatry 21: 370-380, 1980.
17. American Psychiatric Association: Diagnostic and Statistical Manual of Mental Disorders (Third Edition). APA, Washington DC, 1980.
18. Sonners AA: 'Negative symptoms': Conceptual and methodological problems. Schizophr Bull, 11: 364-379, 1985.
19. Mayer M, Alpert M, Stastny P et al: Multiple contributions to clinical presentation of flat affect in schizophrenia. Schizophr Bull, 11: 420-426, 1985.
20. Helmchen H, Hippius H: Depressive syndrome im verlauf neuroleptischer therapie. Nervenarzt, 38: 455-458, 1967.
21. Roth S: The seemingly ubiquitous depression following acute schizophrenic episodes: a neglected area of clinical discussion. Am J Psychiatry 127: 91-98, 1970.
22. Shanfield S, Tucker GJ, Harrow M et al: The schizophrenic patient and depressive symptomatology. J Nerv Ment Dis, 151: 203-210, 1970.
23. McGlashan TH, Carpenter WT Jr: Postpsychotic depression in schizophrenia. Arch Gen Psychiatry 33: 231-239, 1976a.
24. McGlashan TH, Carpenter WT Jr: An investigation of the post-psychotic depressive syndrome. Am J Psychiatry, 133: 14-19, 1976b.
25. Weissman MM, Pottenger M, Kleber H et al: Symptom patterns in primary and secondary depression. Arch Gen Psychiatry 34: 854-862, 1977.

26. Cheadle AJ, Freeman HL, Korer, J: Chronic schizophrenic patients in the community. Br J Psychiatry 132: 221-227, 1978.
27. Knights A, Okasha MS, Salih MA et al: Depressive and extrapyramidal symptoms and clinical effects: A trial of fluphenazine versus flupenthixol in maintenance of schizophrenic out-patients. Br J Psychiatry 135: 515-523, 1979.
28. Siris SG, Harmon GK, Endicott J: Postpsychotic depressive symptoms in hospitalised schizophrenic patients. Arch Gen Psychiatry 38: 1122-1123, 1981.
29. Mandel MR, Severe JB, Schooler NR et al: Development and prediction of postpsychotic depression in neuroleptic-treated schizophrenics. Arch Gen Psychiatry 39: 197-203, 1982.
30. Guze SB, Cloninger R, Martin RL et al: A follow-up and family study of schizophrenia. Arch Gen Psychiatry 40: 1273-1276, 1983.
31. Curson DA, Barnes TRE, Bamber RW et al: Long-term depot maintenance of chronic schizophrenic out-patients II: The incidence of compliance problems, side effects, neurotic symptoms and depression. Br J Psychiatry 146: 469-474, 1985.
32. Herz M: Prodromal symptoms and prevention of relapse in schizophrenia. J Clin Psychiatry 46: 22-25, 1985.
33. Planansky K, Johnston R: Depressive syndrome in schizophrenia. Acta Psychiat Scand 57: 207-218, 1978.
34. Roy A: Depression in chronic paranoid schizophrenia. Br J Psychiatry 137: 138-139, 1980.
35. Johnston DAW: Studies of depressive symptoms in schizophrenia I: The prevalence of depression and its possible causes. Br J Psychiatry 139: 89-93, 1981.
36. Strain F, Heger R, Klicpera C: The time structure of depressive mood in schizophrenic patients. Acta Psychiat Scand 65: 66-73, 1982.
37. Moller HJ, von Zerssen D: Depression states occurring during the neuroleptic treatment of schizophrenia. Schizophr Bull 8: 109-117, 1982.
38. Lerner Y, Moscovich D: Depressive symptoms in acute schizophrenic hospitalised patients. J Clin Psychiatry 46: 483-484, 1985.
39. Martin RL, Cloninger R, Guze SB et al: Frequency and differential diagnosis of depressive syndromes in schizophrenia. J Clin Psychiatry, 46: 9-13, 1985.
40. Drake RE, Cotton PG: Depression, hopelessness and suicide in chronic schizophrenia. Br J Psychiatry 148: 554-559, 1986.
41. Elk R, Dickman BJ, Teggin AF: Depression in schizophrenia: a study of prevalence and treatment. Br J Psychiatry 149: 228-229, 1986.
42. Becker RE, Colliver JA, Verhulst SJ: Diagnosis of secondary depression in schizophrenia. J Clin Psychiatry 46: 4-8, 1985.
43. American Psychiatric Association: Diagnostic and Statistical Manual of Mental Disorders (Third Edition-Revised). APA, Washington DC, 1987.
44. Prosser ES, Csernansky JG, Kaplan J et al: Depression, Parkinsonian symptoms and negative symptoms in schizophrenics treated with neuroleptics. J Nerv Ment Dis 175: 100-105, 1987.

45. De alarcon R, Carney MWP: Severe depressive mood changes following slow-release intramuscular fluphenazine injection. Br Med J 3: 564-567, 1969.
46. Floru, L, Heinrich K, Witter F: The problem of post-psychotic schizophrenic depressions and their pharmacological induction. Int Pharmacopsychiatry, 10: 230-239, 1975.
47. Knights A, Hirsch SR: 'Revealed' depression and drug treatment for schizophrenia. Arch Gen Psychiatry 38: 806-811, 1981.
48. Hirsch SR: Depression 'revealed' in schizophrenia. Br J Psychiatry 140: 421-424, 1982.
49. Galdi J: The causality of depression in schizophrenia. Br J Psychiatry 142: 621-625, 1983.
50. Van Putten T, May PRA: 'Akinetic depression' in schizophrenia. Arch Gen Psychiatry 35: 1101-1107, 1978.

DEPRESSION AND PARANOID SCHIZOPHRENIA

David M. Romney

Department of Educational Psychology
University of Calgary, Calgary Alberta

Paranoid Spectrum Disorders

This paper deals with the relationship between psychotic depression and paranoid schizophrenia as one of the paranoid spectrum disorders. Although paranoid schizophrenia is still officially recognized as a subtype of schizophrenia, there is considerable evidence to suggest that paranoid schizophrenia differs from other forms of schizophrenia (1). Thus, paranoid schizophrenics tend to be older on first admission to hospital and their prognosis is generally better. They also differ markedly from nonparanoid schizophrenics on a variety of cognitive tasks where their performance is characterized by rigidity and by a hypersensitivity to stimuli. For instance, they resist shifting from one concept to another and demonstrate perceptual overconstancy. It is noteworthy that the performance of normal subjects on these tasks falls between that of paranoids and nonparanoids, thereby highlighting the paranoid-nonparanoid dichotomy. Numerous authorities (2,3) have maintained that the paranoid syndrome should be made independent of schizophrenia. The evidence from familial studies provides further support for this distinction. Kendler and Davis (4) concluded that there is an indication that schizophrenia is rarer amongst the relatives of paranoid schizophrenics compared with nonparanoid schizophrenics and that paranoid schizophrenia breeds true to type, i.e., the close relatives of paranoid schizophrenics are more likely to be paranoid than nonparanoid, though this latter finding is debatable (5).

Elsewhere I have argued (6) that paranoid schizophrenia should be regarded as the most pathological stage in the paranoid process. I believe this process is progressive though its course is not inevitable; a patient's condition may well be arrested at an intermediate stage. The paranoid process as a dimension of severity is shown in Figure 1. In keeping with this formulation of the paranoid process, it has in fact been observed that a significant number of cases of

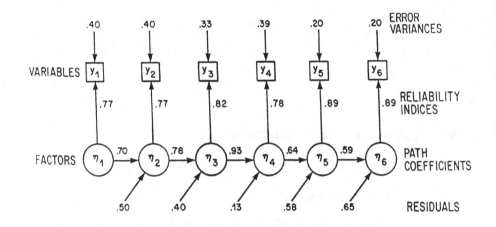

Figure 1   A Six Variable Simplex Model of the Paranoid
            Process (6)

paranoia deteriorate to paranoid schizophrenia (7).

Throughout the rest of this paper, I shall use the terms
paranoia, paranoid disorder, and paranoid schizophrenia
synonymously.

Differential Diagnosis of Paranoid Disorder and Psychotic Depression

In DSM-III-R (8) the differential diagnosis between
paranoid disorder and psychotic depression can be difficult
because depressed patients often manifest paranoid symptoms
and paranoid patients are often depressed (9). Consequently,
the differential diagnosis may depend largely on whether the
delusional or affective symptoms are detected first.
However, one would expect the depressive symptoms to be more
severe and more stable in psychotic depression than in
paranoid disorder, and similarly with respect to the paranoid
symptoms in paranoid disorder. An interesting example of
this diagnostic conundrum was posed by a case described by
Ward et al (10). Their patient was first diagnosed as
suffering from late onset paraphrenia but subsequently
depressive symptoms became paramount. After unsuccessful
experimentation with drug therapy, she was found to respond
positively to the dexamethasone suppression test and given
ECT which alleviated both her depressive and paranoid
symptomatology. What then was her "true" diagnosis? Here is
a case where the depressive symptoms revealed themselves
after the paranoid symptoms but nevertheless appeared primary
and responded to anti-depressive therapy.

The Paranoid-Depressive Continuum

One way to resolve the confusion surrounding the differential diagnosis of paranoid disorder and psychotic depression is to hypothesize a paranoid-depressive continuum. The notion that paranoia and depression lie on a continuum is not new. In 1964 Schwartz (11) published a paper entitled "The Paranoid-Depressive Existential Continuum" in which he conceptualized the continuum as one of personal responsibility: depressives blame themselves, paranoids blame others. Salzman (12), a few years earlier, put forward the view that paranoia was a denial of low self-esteem, the latter being correlated with depression. Going a step further, Allen (13) emphasized that the two states may substitute for one other and that depression is often found to underlie paranoia.

Figure 2. Hypothesized Bipolar Paranoid-Depression Continuum
and Corresponding Delusions

Mood-Congruent   Mood-Incongruent

DEPRESSIVE   DELUSIONS

Somatic   Persecutory   Jealousy   Erotomania   Grandiose   Bizarre

PARANOID   DELUSIONS

PSYCHOTIC DEPRESSION   PARANOID SCHIZOPHRENIA

Delusions are prominent in both paranoid disorder and psychotic depression. When we consider the way these delusions are classified according to their content, we may be struck by the fact that they can also be ordered along a continuum which parallels the paranoid-depression continuum. This parallel continuum would be anchored at the paranoid end by bizarre or grandiose delusions and at the depressive end by mood-congruent delusions. Inspection of Figure 2 shows how the delusions change as we move away from either end of the continuum where patients would, according to this scheme, be less paranoid and more depressed or less depressed and more paranoid, depending on which end we start from. In particular, the mood-incongruent delusions of thought insertion, thought broadcasting and of being mysteriously controlled would be more towards the paranoid end of the continuum. These kinds of delusions are less common in depressed patients (DSM-III-R, p. 220) and one cannot help but wonder if the true diagnosis is not paranoid disorder with depression. Somatic delusions of cancer or any other serious illness (real or imaginary) which occur in depressives may be indistinguishable from the somatic delusions which occur in paranoid patients. Freud's Wolf Man, who had a history of depression serious enough to lead to his hospitalization, subsequently manifested hypochondriacal delusions about his nose and believed that everybody was staring at it (14). This delusion, together with other signs, was interpreted by Freud as indicative of paranoia. This classic case therefore represents an individual whose paranoid delusions could be placed at the depressive end of the paranoid-depressive delusional continuum. In addition, the Wolf Man illustrates how depression can change with time to paranoia. Turning to more contemporary studies, there have been several longitudinal studies on samples of patients which have shown that over the course of time as depression decreases paranoia increases (15,16), or that mood-incongruent delusions supplant mood-congruent delusions as the depression diminishes (17). One would also expect to find instances, such as the Ward et al (10) case study already cited, which will show paranoia giving way to depression. The significance of having a possible trend in both directions will become clear when we later examine the theories which have been proposed to explain the phenomena.

There have been a number of studies using psychological tests which have shown that formal thought disorder, previously considered to be unique to schizophrenia, also appears in patients diagnosed as depressed (18,19) or manic (20,21). For instance, Carter (19, p. 336) found that "psychotic depressives show thought deficits in the same areas as chronic paranoid schizophrenics." Furthermore, in a reply to a criticism of her study on methodological grounds (22), she adds (23, p. 346) that "[paranoid schizophrenics] share symptoms with unipolar psychotic depression both in the area of thought disorder and depressive distortions, a symptom cluster identified with depression." Clearly her findings are consistent with the notion of a paranoid-depressive continuum.

If paranoid and depressive disorders are linked, we might expect a greater tendency for the close relatives of paranoid schizophrenics than for the close relatives of non paranoid schizophrenics to exhibit disorders of mood. With regard to this point, there is some evidence of a higher risk of mood disorders amongst the relatives of paranoid schizophrenics than amongst the relatives of nonschizophrenics (24), but the evidence is conflicting (5). In part these discrepant findings may be due to the different criteria which have been used for diagnosing paranoid schizophrenia and to the different algorithms which have been used to calculate morbidity risk. Future studies with standardized procedures are needed to settle these issues.

Theoretical Explanations

How do individuals become depressed or paranoid? Everyone agrees that environmental stress can play a role in their causation. A person who continually meets with failure and rejection can either blame himself or others for his misfortunes. If he chooses to blame himself he will experience a precipitous drop in self-esteem and become depressed. By blaming others he manages to preserve his self-esteem but simultaneously sows the seeds of paranoia. Attribution theory has been invoked to explain depression (25,26), i.e., the depressed patient attributes bad events to himself and good events to others or to luck. Many studies, some using the Attributional Style Questionnaire (27,28) specially devised for this purpose, have confirmed this finding, though admittedly most of these studies have been analogue studies on university students. We have just completed an investigation of the attributional style of paranoid patients to determine if the reverse is true, i.e., they should attribute good events to themselves and bad events to others or bad luck. Our preliminary findings based on 44 cases confirm this hypothesis. Similarly, Zimmerman et al (29), using the ASQ, found that depressives could not be discriminated from schizophrenics by their attributions of negative events. But it should be noted that the schizophrenics in this study were not exclusively paranoid, and not all the depressed patients were psychotic.

The most popular theory of paranoia is psychodynamic. Those who adhere to this theory believe that paranoia is a defence against the painful feelings of depression. Instead of directing recriminatory thoughts inwards, paranoids direct them outwards. In other words, the defence mechanism of projection protects the individual from low self-esteem and depression. The hypothesis that paranoia shields one from low self-esteem and depression is very persuasive and is consistent with the laboratory findings of Heilbrun and Bronson (30) who were able to induce paranoid thinking in normals as a response to negative evaluation. If the defence mechanism works well the depression will be completely masked; however, if the defence mechanism works only partially the depression will still manifest itself to some degree. Whereabouts the disorder will lie on the paranoid-depressive continuum will depend therefore on the extent to which the defence mechanism of projection is utilized: no utilization will result in depression, whereas full utilization will result in paranoid disorder.

The most comprehensive version of this psychodynamic theory of paranoia has been elaborated by Zigler and Glick (1,31). They point out that projection is not the only defence mechanism used against depression, and that denial, which can give rise to mania, has traditionally been invoked by psychodynamic theorists. It is alleged that the function of the euphoria which typifies mania is to counteract dysphoria and that flight of ideas prevents depressive ruminations (32). Both mania and paranoia then are at opposite poles to depression. In fact mania and paranoid schizophrenia have much in common; for instance, grandiose delusions, as well as irritability and anger (DSM-III-R, p.

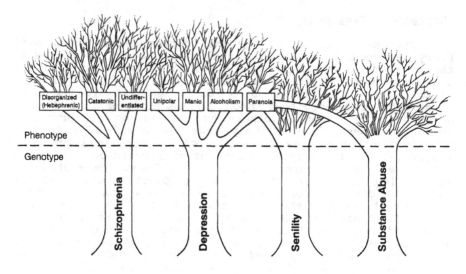

Figure 3.   Conceptual Model (1)

217), and may be almost impossible to differentiate when the mania is severe (15).

What makes Zigler and Glick's contribution original is their metaphorical conceptualization of depression and schizophrenia as genotypes as opposed to paranoia and mania which are phenotypes, i.e., external manifestations of the genotypes.   They argue that a single genotype can produce several different phenotypes and, conversely, that a single phenotype may be determined by more than one genotype. Paranoia and mania are both phenotypes of depression but so, they contend, is alcoholism.   Moreover, paranoia can be caused by dementia and substance-abuse as well as depression (see Figure 3).

Onset of Depression in Paranoid Schizophrenics

Currently there is an argument about whether depression occurs during the early, acute phase of schizophrenia (33) or during the later (postpsychotic) phase (34).   Both findings are consistent with Zigler and Glick's formulation.   In paranoid schizophrenia, one would expect "depressive symptomatology to be manifested either in the early stages of the formation of delusions or at the time when paranoid symptomatology is abating" (35, pp. 161-162).   In other words, there should be an inverse relationship between depressive and paranoid symptomatology during the course of the disorder.   This corollary has so far been borne out by the preliminary results from our own study.

CONCLUSIONS

There is an accumulation of evidence from a variety of sources that (a) paranoid schizophrenia is different for nonparanoid schizophrenia and that (b) paranoid schizophrenia and psychotic depression are at the opposite ends of a bipolar continuum. However, the evidence for the existence of this continuum is still largely circumstantial and a lot more empirical research work needs to be done to test this hypothesis before the diagnosis of paranoid-depressive psychosis can be given serious credence. In future studies of depression in schizophrenia I suggest that paranoid schizophrenics be identified and analyzed as a separate group. Meanwhile, another possibility to be considered is the existence of a single psychoticism factor which predisposes one to psychosis in general (19). Such a factor could account for the appearance of two ostensibly distinct psychotic syndromes in the same individual.

REFERENCES

1. Zigler E, Glick M: Paranoid schizophrenia: An unorthodox view. Am J Orthopsychiatry 54: 43-70, 1984.
2. Magaro P: Cognition in Schizophrenia and Paranoia. Hillsdale, NJ, Erlbaum, 1980.
3. Meissner W: The Paranoid Process. New York, Aronson, 1978.
4. Kendler K, Davis K: The genetics and biochemistry of paranoid schizophrenia and other paranoid psychoses. Schizphr Bull 7: 689-709, 1981.
5. Tsuang MT, Winokur G, Crowe R: Morbidity risks of schizophrenia and affective disorders among first degree relatives of patients with schizophrenia, mania, depression and surgical conditions. Br J Psychiatry 137: 497-504, 1980.
6. Romney D: A simplex model of the paranoid process: Implications for diagnosis and prognosis. Acta Psychiat Scand 75: 651-655, 1987.
7. Munro A: Delusional (paranoid) disorders: Etiologic and taxonomic considerations. II. A possible relationship between delusional and affective disorders. Can J Psychiatry 33: 175-178, 1988.
8. American Psychiatric Association. Diagnostic and Statistical Manual of Mental Disorders (3rd ed rev). Washington, DC, American Psychiatric Press.
9. Roy A: Depression in chronic paranoid schizophrenia. Br J Psychiatry 137: 138-139, 1980.
10. Ward NG, Strauss MM, Ries R: The dexamethasone suppression test as a diagnostic aid in late onset paranoia. J Nerv Ment Dis 170: 240-250, 1982.
11. Schwarz DA: A review of the "paranoid" concept. Arch Gen Psychiatry 8: 349-361, 1963.
12. Salzman L: Paranoid states: Theory and therapy. Arch Gen Psychiatry 2: 679-693, 1960.
13. Allen TE: Suicidal impulse in depression and paranoia. Int J Psychoanal 48: 433-438, 1967.
14. Frosch J: Severe regressive states during analysis: Introduction. J Am Psychoanal Ass 15: 491-507, 1967.

15. Carlson G, Goodwin F: The stages of mania: A longitudinal analysis of the manic episode. Arch Gen Psychiatry 28: 221-228, 1973.
16. Kovacs M, Beck A: Maladaptive cognitive structures in depression. Am J Psychiatry 135: 525-533, 1978.
17. Kettering RL, Harrow M, Grossman L, Meltzer HY: The prognostic relevance of delusions in depression: A follow-up study. Am J Psychiatry 144: 1154-1160, 1987.
18. Payne RW, Hirst H: Overinclusive thinking in a depressive and a control group. J Consult Psychol 21: 186-188, 1957.
19. Carter ML: The assessment of thought deficit in psychotic unipolar depression and chronic paranoid schizophrenia. J Nerv Ment Dis 174: 336-341, 1986.
20. Andreasen NC, Powers PS: Overinclusive thinking in mania and schizophrenia. Br J Psychiatry 125: 452-456, 1974.
21. Harrow M, Grossman L, Silverstein M, Meltzer MD: Thought pathology in manic and schizophrenic patients. Arch Gen Psychiatry 39: 655-671, 1982.
22. Kay SR: Thought deficit in psychotic depression and chronic paranoid schizophrenia: Methodological and conceptual issues. J Nerv Ment Dis 174: 342-344, 1986.
23. Carter ML: Phenotype vs. genotype: A reply to Stanley R. Kay. J Nerv Ment Dis 174: 345-347, 1986.
24. Pope HG Jr, Lipinski JF. Diagnosis in schizophrenia and manic-depressive illness. Arch Gen Psychiatry 35: 811-828, 1978.
25. Beck A: Depression: Clinical, Experimental and Theoretical Aspects. New York, Harper and Row, 1967,
26. Seligman MEP: Helplessness: On Depression, Development, and Death. San Francisco, Freeman.
27. Peterson C, Semmel A, von Baeyer C, Abramson LY, Metalsky GI, Seligman MEP: The Attributional Style Questionnaire. Cog Ther Res 6: 287-299, 1982.
28. Peterson S, Villanova P: An expanded Attributional Style Questionnaire. J Abnorm Psychol 97: 87-89, 1988.
29. Zimmerman M, Coryell W, Corenthal C, Wilson S: Dysfunctional attitudes and attribution style in healthy controls and patients with schizophrenia, psychotic depression, and nonpsychotic depression. J Abnorm Psychol 95: 403-405, 1986.
30. Heilbrun AB, Bronson N: Fabrication of delusional thinking in normals. J Abnorm Psychol 84: 422-425, 1975.
31. Zigler E, Glick M: Is paranoid schizophrenia really camouflaged depression? Am Psychol 43: 284-290, 1988.
32. Arieti SA, Bemporad J: Severe and Mild Depression. New York, Basic Books, 1978.
33. Hirsch SR: Depression "revealed" in schizophrenia. Br J Psychiatry 140: 421-423, 1987.
34. Siris SG, Rifkin A, Reardon GT, Endicott J, Pereira DH, Hayes R, Casey E: Course-related depressive syndromes in schizophrenia. Am J Psychiatry 141: 1254-1257, 1984.
35. Zigler E, Glick M: A Developmental Approach to Adult Psychopathology. Wiley, New York, 1986.

THE DEVELOPMENT OF METHODS TO ASSESS THE TEMPORAL RELATIONSHIP OF DEPRESSIVE AND PSYCHOTIC SYMPTOMS IN SCHIZOPHRENIA

Michael Foster Green, Keith H. Nuechterlein, Jim Mintz and Joseph Ventura

Department of Psychiatry and Biobehavioral Sciences, University of California, Los Angeles

Depression presents a serious problem for many schizophrenic patients. However, little is known about the nature and course of this depression. Van Putten and May (1) suggested that some of the observed depression in schizophrenia may be a drug-induced akinesia syndrome. It has been proposed that depression in schizophrenia can be a "reactive" depression to the disruptive effects of the psychotic episode (2). McGlashan (3) introduced the term "aphanisis" to describe the residual schizophrenic defect state that might be confused with depression. It has also been argued that postpsychotic depression actually begins concurrent with psychotic symptoms but that the depressive symptoms are not noticeable until the florid symptoms begin to remit (4,5). In addition, Herz and Melville (6) demonstrated that depression is commonly reported in the prodromal period, suggesting that "prepsychotic" or prodromal depression may be an addition relevant type of depression.

We might expect depression to be more prevalent in a group of schizophrenic patients than in a nonpsychiatric group. However, if depression in schizophrenia is related to periods of psychosis as some of the conceptions mentioned above would imply, we might expect to see a systematic time relationship between the onset of depression and the onset of psychosis. Prospective ratings of depressive and psychotic symptoms across multiple episodes would allow one to determine if the onset of depression occurs disproportionately in the time periods just before (prodromal), concurrent with, or after the onset of a psychotic episode, or after the offset of psychosis (postpsychotic). Alternatively, depressive symptoms might not show any temporal relationship to the psychotic episode onset.

The present report describes recent efforts to consider the course of depressive and psychotic symptoms from a longitudinal study of recent-onset schizophrenic patients. Prospective symptom ratings were made every two weeks for a period of at least 1 year. The data were treated in two different ways: first, the ratings were treated as _categorical_ variables to consider the temporal relationship of clinically meaningful psychotic and depressive episodes and to clarify empirically whether a pattern of prodromal, concurrent, or postpsychotic depression can be demonstrated in the early phase of schizophrenia (7). Second, we will report on a work in progress that uses time series analyses and treats the data as _continuous_ variables. For this section, we will discuss the time series method and present an illustrative case.

METHOD-CATEGORICAL ANALYSES

Subjects

Subjects were participants in an ongoing longitudinal study of the early phase of schizophrenia: Developmental Processes in Schizophrenic Disorders. This project examines the nature and determinants of the course of illness for at least a 3-year period. Subjects are selected for the project if they: a) meet criteria for schizophrenia or schizoaffective disorder, mainly schizophrenic, according to Research Diagnostic Criteria (RDC), b) have had their first psychotic episode within 2 years of evaluation, c) are 18 to 25 years of age, and d) are Caucasian or acculturated Hispanic or Asian. All diagnoses are based on an expanded version of the Present State Exam and all interviewers are trained to at least an 85% minimum agreement on symptom presence with the Diagnosis and Psychopathology Unit of the Clinical Research Center for the Study of Schizophrenia at UCLA (Robert P. Liberman, PI). Subjects are excluded for habitual drug or alcohol abuse, evidence of a CNS disorder, or mental retardation. For the current study, we selected subjects (n = 27) who showed some features of both depressive and psychotic symptoms during the follow-up period.

Symptom Ratings

Symptom ratings were made on an expanded version of the Brief Psychiatric Rating Scale (BPRS; 8) which includes behavioral anchors (9) for the 7-point scale. The ratings were made every 2 weeks by clinicians trained to a criterion reliability level (median intraclass correlation of at least .80). The critical scales used for the psychotic symptoms were: Unusual Thought Content, Hallucinations, and Conceptual Disorganization. The scales for the depressive symptoms included: Depression, Guilt, and Suicidality. When subjected to factor analysis, the depressive symptoms do not cluster with the negative symptoms on the BPRS (i.e., Blunted Affect, Motor Retardation, and Emotional Withdrawal), but rather represent an independent group of symptoms.

Relapse Criteria

We used previously developed BPRS criteria (10) for the onset of a psychotic relapse or psychotic significant exacerbation. The following are brief definitions of the relapse criteria. The full criteria are listed elsewhere (10).

Psychotic relapse involves a severe (six) or very severe (seven) rating on any one of the three psychotic scales after at least a 1 month remission.

Remission is defined as a period of at least 1 month when the patient maintained a score of three or less on all psychotic scales.

Psychotic significant exacerbation involves a moderately severe (five) rating on any of the psychotic scales plus an increase of at least two points on one of the other psychotic scales after remission.

Psychotic persisting symptoms-significant exacerbation is rated when the subject continues to have a score of 4 or more on one of the psychotic scales and then shows at least a 2-point increase to a (six) or (seven).

Following the previously developed BPRS criteria (9), we defined the onset of a depressive episode in a way that paralleled the definition of psychotic onset. The only difference was the inclusion of the category of mild depressive exacerbation to allow for the detection of low levels of depressive symptoms.

Depressive relapse involves a severe (six) or very severe (seven) rating on any of the three depression scales after at least a 1 month depressive remission (scores of three or below on all three depression scales).

Depressive significant exacerbation involves a moderately severe (five) rating on any depressive scale plus an increase of 2 points on another of the scales following depressive remission.

Depressive persisting symptoms-significant exacerbation is rated when the patient continues to have a score of four or above on one of the depressive scales and then shows a 2-point increase to a score of six or seven.

Depressive mild exacerbation is rated when a moderate (four) rating is obtained on one of the depressive scales for at least 1 month following a depressive remission.

Time Periods

Six discrete time periods were defined, using the onset of the psychotic episode as a reference point.

1) Prodromal: the 8 week period prior to the onset of a psychotic episode.
2) Concurrent: the 2-week BPRS period during which the psychotic symptoms reached criteria for a psychotic episode.
3) Soon after psychotic onset: the first 4 weeks after onset of psychosis.
4) Long after psychotic onset: period from the end of the first 4 weeks after psychotic onset until the patient

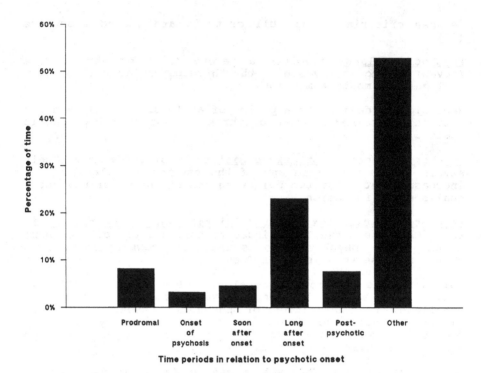

Figure 1. Phases of Psychotic Episodes

enters psychotic remission.
5) <u>Postpsychotic</u>: the 8-week period after the end of a
    psychotic episode.
6) <u>Other</u>: any remaining periods during which the BPRS
    ratings were obtained.

RESULTS: CATEGORICAL ANALYSES

     BPRS ratings were obtained on the 27 subjects for a mean
of 153 weeks (range = 57-297 weeks). Figure 1 shows the
proportion of time spent in each of the 6 time periods.
During the follow-up period, patients spent over half their
time (54.8%) in a nonpsychotic period that was not near a
psychotic episode and (23.6%) of their time in a psychotic
period more than a month after onset of psychotic episode.
The remaining time was divided as follows: 7.4% for
prodromal, 3.0% for concurrent, 4.2% for soon after, and 7.0%
for postpsychotic.

     A total of 63 depressive onsets were recorded for all
subjects over the 6 time periods. The mean number of

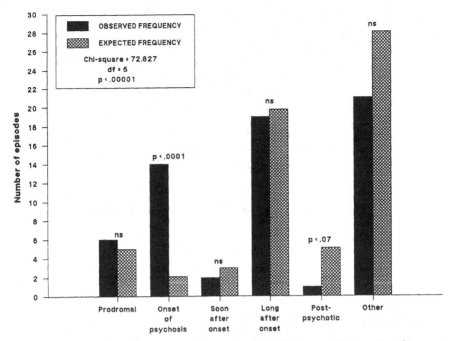

Figure 2.   Onset of Depression During Phases of Psychotic
            Episodes

depressive periods across time periods was 2.33 (sd = 1.27)
for each subject. Most subjects (18 out of 27) showed more
than one depressive episode.   Of these 18 subjects, only 3
had all of their depressive episodes during the same type of
time period. Two of these subjects had all of their episodes
in the "long after" period, and 1 had the episodes in the
"other" period.

        An expected frequency of depressive episodes for
each subject was determine for each of the time periods.
This expected frequency was based on the relative duration of
the periods and the subject's total number of depressive
periods.   Next, the expected frequencies were summed across
subjects to yield an overall expected frequency for each time
period.   Figure 2 compares the expected frequencies of the
depressive onsets to the observed frequencies. As seen here,
the difference in the distributions is highly significant,
Chi square = 72.8 p < .00001.

        When we examine the time periods separately, we see that
the onset of depression was found to be concurrent with
psychotic onset to a highly disproportionate degree, (p <
.0001).   While less than 2 depressive onsets would be
expected during this period, 14 were found.   The
postpsychotic period showed a nonsignificant tendency to have
too _few_ depressive episodes (p < .07). The results from the
remaining categories did not differ from the expected
frequencies.

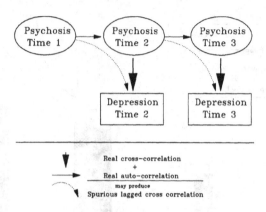

Figure 3. A Schematic of Autocorrelation

METHOD: TIME SERIES ANALYSIS

Treating the data as categorical variables seems logical from a clinical perspective because the focus is on symptoms which have met a certain threshold. Treating data in a continuous fashion as in time series analyses has some advantages. First, the categorical analyses focus on depressive onset and ignore offset. Time series analyses give weight to changes in either direction. Second, categorical analyses examine only those depressive symptoms which have reached a predetermined level of clinical significance. Time series analyses would also take into account sub-clinical changes in symptoms. Third, time series models identify relatively fixed time-locked relationships between symptoms, and are less forgiving than categorical analyses if such relationships vary within individual.

In time series analyses, we can consider the cross-correlations between the psychotic index (sum of the 3 psychotic items) and depressive index (sum of the 3 depressive items). A concurrent-cross correlation would use the symptom data taken from the same BPRS. However, it is also possible to do a lagged correlation in which the depressive index from one BPRS is correlated with the psychotic index from a following or preceding BPRS. Using this method we can obtain a series of correlations which differ in the amount of lag between the two scales.

There is a potential problem with this method: autocorrelation might yield spurious results. Autocorrelation is the correlation of an index with itself across repeated measurements. Hence, if the psychotic scores from one BPRS correlate significantly with the psychotic scores from the following BPRS (which they often do), a potentially misleading autocorrelation is present. The problem with autocorrelation is diagrammed in Figure 3.

Let us assume that we have a concurrent relationship between depressive and psychotic symptoms, but no lagged correlations. For example, the depressive index could be correlated with the psychotic index from the same BPRS, but not correlated with the psychotic index from the preceding BPRS. If the psychotic index from one BPRS correlates with the psychotic index from the next BPRS (autocorrelation, which is indicated by the solid horizontal arrows), then a spurious correlation (indicated by the dotted arrows) will be observed between psychosis from the previous BPRS and current depression. This spurious correlation only occurs because psychosis at time 1 is related to psychosis at time 2.

To avoid the problems associated with autocorrelation, both the depressive and psychotic indexes should be converted by a transfer function that eliminates the autocorrelation. By transforming the data in this fashion, we can eliminate the spurious correlations caused by the autocorrelation, and lagged correlations can be interpreted more clearly.

An additional type of time series analyses is spectral analysis. Spectral analysis tries to determine the composition of a complex wave form by fitting sine waves to the data. Such a procedure is regularly conducted to identify the composition of EEG recordings. The changes over time in the psychotic or depressive indexes can also be considered as complex wave forms. The spectral analyses generate a series of overlapping sine waves which, when combined, approximate the psychotic (or depressive) symptom patterns over time. Figure 4 shows how two sine waves can be combined to yield a composite wave.

RESULTS: TIME SERIES ANALYSES

1)   Lagged correlations

We have selected case #138 as an example because this subject shows considerable variability in both psychotic and depressive symptoms and because data are available for a 5 year period. Although the BPRS ratings were collected every 2 weeks, the BPRS data were collapsed into periods of one month to minimize the effects of any missing interviews. Table 1 shows the lagged correlations between the psychotic and depressive symptoms. The first column shows the correlations between the psychotic and depressive symptoms. The first column shows the correlations without taking autocorrelation into account. This procedure yields a strong correlation between psychotic symptoms and the concurrent level of depressive symptoms (r = .47). In addition, psychotic symptoms show a small correlation with the depressive symptoms from the preceding (r = .14) and following (r = .21) BPRS interviews.

The second column shows the lagged correlations after removing the effects of autocorrelation. In this case, the concurrent correlation remains high (r = .45) but the correlations from the immediately preceding (r = .01) and following (r = -/09) BPRS interviews are smaller than in the first column. Hence, it appears that some of the lagged correlations in the first analyses were slightly inflated.

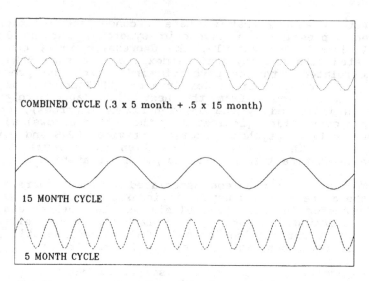

COMBINED CYCLE (.3 x 5 month + .5 x 15 month)

15 MONTH CYCLE

5 MONTH CYCLE

Figure 4.   Frequency Domain Analysis (Adding Two Sine Waves)

Table 1.   Lagged Correlations for Subject #138

		Original Data	Transformed Data
Depression leads Psychotic Symptoms	-3	.16	-.13
	-2	.22	.15
	-1	.14	.01
Current	0	.47	.45
Depression follows Psychotic Symptoms	+1	.20	-.09
	+2	.16	.15
	+3	-.07	.01

2) Spectral Analyses

The psychotic symptoms for subject #138 are diagrammed in Figure 5. Over the 5 year period, the subject shows a large amount of variability in psychotic symptoms and the pattern appears very complex. Spectral analyses revealed prominent components consisting of a 5-month cycle and a 15-month cycle. In Figure 6, a composite wave of 5-month and 15-month cycles is graphed with the data, and we see that the fit is surprisingly good (R = .55). Hence, the pattern of psychotic symptoms, which initially looked almost random, is closely described by 2 overlapping cycles: one of 5 months and one of 15 months. The same type of spectral analyses could be conducted on the depressive symptoms or any other longitudinal data base.

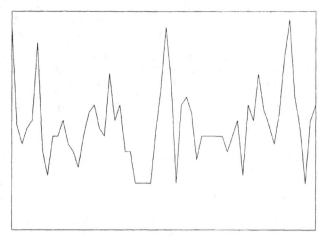

Figure 5.   BPRS Psychosis Cluster (Case 138 over 5 years)

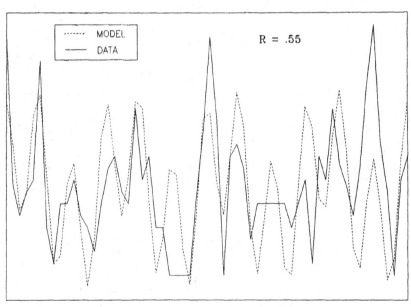

Figure 6.   Frequency Domain Modeling (Combined 5 and 15 Month Cycles)

GENERAL DISCUSSION

The categorical analyses revealed that pathological levels of depressive and psychotic symptoms occurred simultaneously much more frequently than expected by chance. Depressive onset tended to occur less often than expected by chance during the postpsychotic period (defined here as the 8-week period following the end of the psychotic episode). No evidence for an increased frequency of depressive onset during prodromal or ongoing psychotic periods was found. It is possible that some patients show systematic patterns of depressive onset during the psychotic period (e.g., soon after, long after) but that these patterns were washed out by averaging across patients.

The time series analyses allow one to examine the temporal relationship between psychotic and depressive symptom fluctuations which characterize individual patients. For the patient used as an illustration of this approach, these analyses yielded a significant concurrent correlation, but no other time-lagged correlations. The finding for this one patient converges with the categorical analyses across patients. In addition, the spectral analyses indicated that the pattern of psychotic symptoms for this subject was described by overlapping 5- and 15-month cycles. The time series analyses might yield subtypes of patients who show different temporal patterns than the ones revealed for this subject.

In summary, we have introduced 2 types of analyses that can potentially make a strong contribution to our understanding of the temporal course and relationship of psychiatric symptoms. Both types of analyses support the notion that depressive onset occurs concurrent with psychotic onset to a disproportionate degree, but not excessively in the time periods before or after the episodes. Thus, we find no distinctive post psychotic depression linked temporally to psychotic periods in these recent-onset schizophrenic patients. The categorical analyses have the advantage of considering clinically relevant psychotic and depressive episodes, while the time series analyses have the advantage of being able to incorporate subclinical changes and symptom offset as well as onset. One type of time series analysis, spectral analysis, holds particular promise because it can determine the spectral composition of fluctuations in symptoms that would otherwise appear random to the clinician.

FOOTNOTE

The authors would like to thank Sun Hwang, M.S., M.P.H. and Tim Welke for their help with data collection and analysis. This work was supported by NIMH grant MH 37705 to Dr. Nuechterlein. The data analysis, diagnostic training, and symptom assessment training was supported by NIMH Clinical Research Center Grant MH 30911 (Robert P. Liberman, P.I.).

REFERENCES

1. Van Putten T, May PRA: "Akinetic depression" in schizophrenia. <u>Arch Gen Psychiatry</u> 35: 1101-1107, 1978.
2. Hartmann W, Kind JE, Muller P, Steuber H: Neuroleptic drugs and the prevention of relapse in schizophrenia: A workshop report. <u>Schizophr Bull</u> 6: 536-543, 1980.
3. McGlashan TH: Aphanisis: The syndrome of pseudo-depression in chronic schizophrenia. <u>Schizophr Bull</u> 8: 118-134, 1982.
4. Moller H, von Zerssen D: Depressive states occurring during the neuroleptic treatment of schizophrenia. <u>Schizophr Bull</u> 8: 109-117, 1982.
5. McGlashan TH, Carpenter WT: An investigation of the postpsychotic depressive syndrome. <u>Am J Psychiatry</u> 133: 14-18, 1976.
6. Herz MI, Melville C: Relapse in schizophrenia. <u>Am J Psychiatry</u> 137: 801-805, 1980.
7. Green MF, Nuechterlein KH, Ventura J, Mintz J: The temporal relationship between depressive and psychotic symptoms in recent-onset schizophrenia. Submitted for publication.
8. Overall JE, Gorham DR: The Brief Psychiatric Rating Scale. <u>Psych Rep</u> 10: 799-812, 1982.
9. Lukoff D, Nuechterlein KH, Ventura J: Appendix A. Manual for the Expanded Brief Psychiatric Rating Scale (BPRS). <u>Schizophr Bull</u> 12: 594-602, 1986.
10. Nuechterlein K, Miklowitz DJ, Ventura J, Stoddard M, Lukoff D: A system for classifying outcomes in major psychiatric disorders. Manuscript in preparation.

NEGATIVE SYMPTOMS AND DEPRESSION IN SCHIZOPHRENIA

Michael Pogue-Geile

Department of Psychology and Department of
Psychiatry, University of Pittsburgh
Pittsburgh, Pennsylvania

The question of the relationship between negative symptoms and the various signs and symptoms of depression is an important one given the current great interest in both negative and depressive symptoms in schizophrenia and the potential similarity between the two concepts (1). Table 1 highlights this conceptual similarity between common definitions of both negative symptoms and depression. As should be apparent, the behavioral signs that are generally considered as indicators of the negative symptom syndrome partially overlap with the broader group of signs and symptoms that are usually taken as indicative of a depressive syndrome. The potential consequences of this conceptual ambiguity are made more acute by the differences in theoretical hypotheses that are generally held for each syndrome. Table 1 also outlines some of these discrepancies between hypotheses concerning negative and depressive symptoms in schizophrenia. As can be seen, most hypotheses predict that negative symptoms will be associated with more severe characteristics, whereas depressive symptoms are usually predicted to be related to more benign characteristics. Given such disparate hypotheses and yet such similar concepts, it becomes crucial to know to what degree "negative" and "depressive" symptoms may be distinguishable among schizophrenic patients. The aim of the current research was therefore to examine the nature of this relationship among schizophrenic patients between negative symptoms and the various signs and symptoms of depression.

Several studies have investigated aspects of this question, but with varied results and some methodological shortcomings. These studies of the relationship between negative symptoms and depression are of two general types. Seven studies have examined the occurrence of negative symptoms among diagnosed depressed patients in comparison to schizophrenic patients. Three of the five studies that assessed subjects as inpatients reported that rates of negative symptoms were high and not significantly different

Table 1 Negative and Depressive Symptoms: Comparison of Definitions and Hypotheses

Negative Symptoms	Depressive Symptoms

I. Operational Definitions

Core Symptoms	Core Symptoms
Flat Affect	Depressed Mood
Poverty of Speech	Guilt
	Weight Loss or Gain

Other Suggested Symptoms	
	Insomnia or Sleep Excess
	Suicidal Ideation
Psychomotor Retardation	Psychomotor Abnormalities
Fatigue	Fatigue
Withdrawal	Anhedonia
Apathy	Decreased Concentration
Anhedonia	
Distractibility	

II. Hypotheses

Negative Symptoms	Depressive Symptoms
Longitudinally Stable	Longitudinally Unstable
Poor Prognostic Sign	Good Prognostic Sign
Medication Unresponsive	Medication Responsive ?
Cognitive Deficits	No Cognitive Deficits
Structural Abnormalities	Neurotransmitter ?

Note: Depressive signs and symptoms are from the SADS-L (15)

between schizophrenic and depressed groups (2-4), whereas two found higher levels of flat affect among schizophrenics (5,6). In contrast, the two studies that have compared schizophrenic and depressive subjects after hospital discharge found that negative symptoms were less severe among depressed compared to schizophrenic subjects (7,8). These results are consistent with the notion that milder depressive symptoms, such as might be present following hospital discharge, may be observationally distinguishable from negative symptoms, but that signs of severe depression, such as that which leads to inpatient hospitalization, may not be.

A second group of seven studies has investigated the related question of the degree of correlation among schizophrenic patients between depressive and negative symptoms. Six of these studies have found that negative symptoms are not significantly correlated with depressive symptoms in schizophrenia (7,9-13), although one study has found the two significantly positively correlated (14). These mixed findings and the frequent methodological shortcomings of these studies argue for further examination of the question of the relationship between negative and depressive symptoms among schizophrenic patients. The present study was designed to address this general issue utilizing several methodological improvements over earlier studies, such as: assessment of schizophrenic outpatients, blind, reliable ratings of negative symptoms based on video tapes, and a comprehensive assessment of depressive symptoms using both clinical interviews and standard self-report questionnaires. This study sought to address the following specific questions.

1. Is a history of depressive episodes a risk factor for current negative symptoms among schizophrenic outpatients?
2. Is the presence of a current depressive syndrome a risk factor for current negative symptoms among stable schizophrenic outpatients?
3. Are specific current depressive symptoms differentially associated with current negative symptoms?

METHOD

Subjects

Forty-four schizophrenic (40) or schizoaffective (4) subjects were recruited from the outpatient clinics of the Western Psychiatric Institute and Clinic at the University of Pittsburgh. Subjects were selected who met the following screening criteria: chart diagnosis of schizophrenia or schizoaffective disorder, between the ages of 18 and 45, not hospitalized in the past 6 months, judged by their clinician to be stable (i.e., not in the midst of an acute episode and not requiring medication adjustment in the previous 6 weeks), no current severe drug or alcohol abuse, and no history of potential organic complications. Eligible patients who gave informed consent were interviewed using a modified Schedule for Affective Disorders and Schizophrenia - Lifetime Version (SADS-L) (15) by trained interviewers, who were advanced graduate students in clinical psychology. Diagnoses were made according to the Research Diagnostic Criteria (RDC) (16) by a trained diagnostic committee composed of a Ph.D. level clinical psychologist (M.P-G.) and two advanced psychology graduate students. Diagnostic decisions were based on videotapes of the SADS-L interview and hospital charts.

Table 2 presents a detailed clinical and demographic description of this sample. In brief, this was a predominantly young male sample, who were still relatively early in the course of their disorder. All subjects were currently outpatients and all were receiving antipsychotic medications.

Measures

1. Negative Symptoms

This study took a narrow approach to defining the negative symptom syndrome (1,17). Only the signs of poverty of speech and flat affect were rated as negative symptoms. Ratings of these two negative symptoms were made using a

Table 2   Demographic and Clinical Characteristics of Sample

Characteristic	Study Patients (N = 44)	
Sex (Male)	29	(67%)
Education	12.9	(3.0)
Age at 1st Hospitalization	25.0	(4.3)
Current Age	34.8	(6.4)
Number Hospitalizations	3.4	(2.3)
Total Months Hospitalized	6.3	(5.6)

<u>Note</u>:   Means (standard deviations)

revised version of Andreasen's Scale for the Assessment of Negative Symptoms (SANS) (18).   For theoretical reasons, inappropriate affect was not included in rating the affective flattening scale, and poverty of content was not included in rating the alogia scale.   Data presented here are from the six-point global ratings of alogia and affective flattening that incorporated these modifications.

Raters (an advanced graduate student and an undergraduate honors student) were trained to a reliability of at least .90 (intraclass correlation) with the author on a set of 10 test videotapes of non-study patient interviews. The author has extensive experience in rating negative symptoms, and in addition has been trained in using the SANS at Iowa.   All study ratings were then made jointly and independently by the two raters based on videotapes of patient interviews, after which all rater disagreements were resolved by consensus discussion among the author and raters. These negative symptom ratings were made blind to assessments of depressive symptoms.

2.   Depressive Symptoms

A lifetime history of depression was assessed retrospectively based on the SADS-L, using the RDC definition of a depressive episode.   This depressive episode could have occurred either alone or in the context of psychotic symptoms.

Current depressive symptoms were assessed using the modified SADS-L, which also inquired about symptoms in the past month.   The RDC was used to define the presence of a current episode of depression.   These ratings of both lifetime and current depressive symptoms were based on videotapes of the SADS-L interview and were made by an experienced diagnostic committee, who were blind to the negative symptom ratings.

In addition to these ratings of past and current depressive symptoms based on the SADS-L, current depressive symptoms were also assessed using a standard 13-item short form of the Beck Depression Inventory (BDI) (19).   This short form has been found to correlate approximately .90 with the original version of the BDI (20).

RESULTS

Frequency of Negative and Depressive Symptoms

Table 3 presents the frequencies of negative symptoms and current and lifetime depressive symptoms in this sample of schizophrenic outpatients. Forty-five percent of the subjects evidenced moderate to severe ratings of either poverty of speech or flatness of affect. Similarly, 46% of the subjects reported experiencing at some time a depressive syndrome severe enough to meet the RDC definition. This episode occurred during a psychotic episode for 30% and alone for 23% of the patients. In contrast to this relatively frequent lifetime rate of depression, only 9% of the patients were currently experiencing an RDC depression in the last month. This relatively low rate of current depression based on the SADS and RDC was also reflected in a low average score for the self-report BDI. Neither negative nor depressive symptom ratings were significantly correlated with current age, sex, education, or duration of previous hospitalizations. These data indicate the presence of sufficient variability for negative and depressive symptoms and provide a context for interpreting their joint association.

Table 3   Frequencies of Negative and Depressive Symptoms

Characteristic	Frequency
1. Current Negative Symptoms[1] (Moderate to severe ratings)	
Flat Affect	42% (16)
Poverty of Speech	21% (8)
Flat Affect or Poverty of Speech	45% (17)
2. Lifetime Depressive Syndrome (SADS-L with RDC definition)	45% (20)
3. Current Depressive Syndrome (SADS-L with RDC definition)	9% (4)
4. Beck Depression Inventory (Short-form, 39 maximum score)	M = 5.8, SD = 6.4

[1]Negative symptom ratings missing for 6 subjects due to lack of video tapes

125

The associations between current negative symptoms and both lifetime and current depressive symptoms appear in Table 4. The top section of this table indicates that a history of a depressive syndrome was not a significant risk factor for the occurrence of current negative symptoms among these young schizophrenic outpatients. Forty-eight percent of subjects without a reported history of depression showed negative symptoms, as did an approximately equal 41% of those with a history of depression. The picture regarding the association between current depression and negative symptoms is similar. Although infrequent, the presence of a current RDC depressive syndrome did not significantly increase the risk for negative symptoms. This was also reflected in the absence of a significant association between negative symptoms and BDI scores. Similar results were found when males and females were analyzed separately, and when the negative symptoms of poverty of speech and flat affect were individually considered. Overall then these data suggest little if any association between negative and depressive symptom ratings among schizophrenic patients.

In order to evaluate potential differential associations within the depressive syndrome, the overall measures of current depressive symptoms were broken down into the following separate subscales based on their content: 1) ideational symptoms, 2) vegetative symptoms, and 3) ambiguous symptoms that might be labeled as either depressive or negative symptoms. As Table 5 shows, negative symptoms rated by observers using the SANS were not significantly associated with any of these individual depression subscales based on the SADS. Similar findings occurred for parallel subscales of the BDI. These data imply that observer-rated negative symptoms are distinguishable from self-reported ideational and vegetative depressive symptoms. In addition, they suggest a possible disassociation between observer- and self-ratings of negative symptoms.

DISCUSSION

No significant associations were observed between depressive and negative symptoms among these schizophrenic patients. This finding held for: lifetime as well as current depression; clinical interview-based as well as self-report questionnaire measures of depression; and overall as well as individual symptom ratings of depression. It also held across different negative symptoms. These "negative" results should be convincing due to a number of methodological strengths of the study. In particular, ratings were made reliably, blindly, and with multiple, standard methods. Furthermore, the subject sample was clinically stable, early in the course of the disorder, and large enough to provide sufficient statistical power to detect moderate associations. These findings are consistent with those of six previous studies on the same question, but contradict recent results by Kay and Lindenmayer (14), who found a significant positive correlation between negative and depressive symptoms among schizophrenic patients. Overall the weight of the evidence seems to favor the conclusion that narrowly defined negative symptom ratings are statistically

Table 4 Associations between Current Negative Symptoms
and Lifetime and Current Depressive Symptoms

| Depressive Symptom | Overall Negative Symptoms[1] | |
	Absent/Mild	Moderate/Severe
1. Lifetime Depression (SADS-L)		
Absent	52% (11)	48% (10)
Present	59% (10)	41% (7)
	Chi-square (1) = .005, p < .94	
2. Current Depression (SADS-L)		
Absent	53% (18)	47% (16)
Present	75% (3)	25% (1)
	Chi-square (1) = .095, p < .76	
3. Beck Depression Inventory		
Mean (SD)	5.5 (7.2)	5.8 (5.7)
	t (32) = .13, p < .89	

[1]Presence of either poverty of speech or flat affect at
moderate to severe levels.

independent of ratings of subjectively reported depressive
symptoms among young schizophrenic patients.

These observations have several implications. On a
surface, methodological level they should be encouraging in
that they provide evidence that negative and depressive
symptomatology can be distinguished using current rating
techniques. Thus studies of negative or depressive symptom
ratings in schizophrenia appear in fact to be investigating
somewhat different phenomena. Nevertheless, future
investigations of either phenomena would do well to assess
both constructs in order to examine further their
discriminative validity.

A second implication is that contrary to some hypotheses,
depressive symptoms do not appear to decrease the chances of
evidencing negative symptoms. Often it is assumed that
patients rated with negative symptoms do not experience
strong emotions, such as depression, since they appear to
have "flattened affect". These data suggest that this is not
usually the case, since no significant negative correlations

Table 5 Association between Negative Symptoms and
Depression Subscales

Current Depressive Symptoms from SADS-L	Total Negative Symptom Score[1]
1. Ideational Symptoms (Depressed mood, guilt, or suicidal ideation)	
Absent	3.0 (2.6)
Present	2.7 (1.2) t (36) = .34, p < .74
2. Vegetative Symptoms (Appetite or sleep problems)	
Absent	3.0 (2.4)
Present	1.0 (0.0)[2]
3. Ambiguous Symptoms (Lack interest, lack energy, poor concentration, or psychomotor problems)	
Absent	3.1 (2.5)
Present	2.3 (1.3) t (36) = .64, p < .53

[1]Sum of individual scores for poverty of speech and flat affect

[2]T-test not calculated due to absence of variance in one group.

between negative and depressive symptoms were observed. Thus there appears to be a distinction between observed negative behavior, such as poverty of speech and flat affect, and subjective reports of depressed mood.

The finding that observed negative symptom behavior ratings also were independent of subjective reports of "ambiguous symptoms" (i.e., those which some might include in either negative or depressive syndromes) suggests an even further possible disassociation between observed behavior and subjective reports within the negative syndrome construct itself. Thus in this study, patients who were observed to show flattened affect and/or poverty of speech were not especially likely to report feeling without interest or energy. Andreasen and Olsen (21) have reported similar findings. These observations suggest that more attention should be paid to assessing the subjective experience of negative symptoms, and that studies should examine the potentially differential correlates of these subjective and objective negative symptom/sign ratings.

Although negative and depressive symptoms appear to be distinguishable among schizophrenic patients, several studies have reported high rates of negative symptoms among depressed inpatients. How might these data be integrated in an overall model of the relationship between negative and depressive symptoms? One possible hypothetical model would define negative "symptoms" as observed behavioral signs, such as poverty of speech and flat affect, that can arise due to several different underlying pathologies. The two most important such putative pathways to these negative signs would be either: 1) processes underlying certain severe depressive states or 2) other distinct processes that would not be associated with the usual manifestations of depression. In this "two-factor" model, negative and depressive symptoms would be generally independent among schizophrenic patients because negative symptoms in this population would be primarily due to the second underlying factor, which is independent of the manifestations of depression. Furthermore, patients in severe depressive states, such as those found among inpatient depressives, should show high levels of "negative symptoms", which have arisen due to the first putative pathological factor. Such a model would also predict a positive correlation between these "negative symptoms" and measures of the severity of depression among diagnosed depressed patients.

Although tentative, this hypothetical model has the virtue of suggesting further questions for investigation. Perhaps most importantly it highlights the question of why the behavioral signs of severe schizophrenia and severe depression are so observationally similar. Might there perhaps be underlying pathophysiological features at some level that are shared between the two? One of the several possibilities that this line of reasoning would suggest is that trials of antidepressant therapies might be of some use for negative symptoms among schizophrenic patients. Recent tentative support on this point provided by Siris and colleagues (22) suggests the potential usefulness of such hypotheses.

ACKNOWLEDGEMENT

Preparation of this chapter was supported in part by grant NIH BRSG RR 07084-20. The author is grateful to Jennifer Brunke, Ann Garrett, John Hall, and Mary Kelly for assisting in interviewing and tape rating.

REFERENCES

1. Pogue-Geile MF, Zubin J: Negative symptomatology and schizophrenia: A conceptual and empirical review. Int. J of Mental Health 16: 3-45, 1988.
2. Andreasen NC: Affective flattening and the criteria for schizophrenia. Am J Psychiatry 136: 944-947, 1979.
3. Berenbaum SA, Abrams R, Rosenberg S, et al: The nature of emotional blunting: A factor analytic study. Psychiatry Res 20: 57-67, 1987.

4. Kulhara P, Chadda R: A study of negative symptoms in schizophrenia and depression. Compr Psychiatry 28: 229-235, 1987.
5. Boeringa JA, Castellini S: Reliability and validity of emotional blunting as a criterion for diagnosis of schizophrenia. Am J Psychiatry 139: 1131-1135, 1982.
6. Chaturvedi SK, Prasad Rao G, Mathai JP, et al: Negative symptoms in schizophrenia and depression. Indian J Psychiatry 27: 237-241, 1985.
7. Pogue-Geile MF, Harrow M: Negative and positive symptoms in schizophrenia and depression: A follow-up. Schizophr Bull 10: 371-387, 1984.
8. Ragin AB, Pogue-Geile MF, Oltmanns, TF: Poverty of speech in schizophrenia and depression during inpatient and post-hospital periods. Br J Psychiatry (in press).
9. Craig TJ, Richardson MA, Pass R, et al: Measurement of mood and affect in schizophrenic inpatients. Am J Psychiatry 142: 1272-1277, 1985.
10. Iager AC, Kirdh DG, Wyatt RJ: A negative symptom rating scale. Psychiatry Res 16: 27-36, 1985.
11. Kay SR, Opler LA, Fiszbein A: Significance of positive and negative syndromes in chronic schizophrenia. Br J Psychiatry 149: 439-448, 1986.
12. Lewine RRJ, Fogg L, Meltzer HY: Assessment of negative and positive symptoms in schizophrenia. Schizophr Bull 9: 368-376, 1983.
13. Prosser ES, Csernansky JG, Kaplan J, et al: Depression, Parkinsonian symptoms and negative symptoms in schizophrenics treated with neuroleptics. J Nerv Ment Dis 175: 100-105, 1987.
14. Kay SR, Lindenmayer JP: Outcome predictors in acute schizophrenia: Prospective significance of background and clinical dimensions. J Nerv Ment Dis 175: 152-160, 1987.
15. Endicott J, Spitzer RL: A diagnostic interview: The Schedule for Affective Disorders and Schizophrenia. Arch Gen Psychiatry 35: 837-844, 1978.
16. Spitzer RL, Endicott J, Robins E: Research Diagnostic Criteria: Rationale and reliability. Arch Gen Psychiatry 35: 773-782, 1978.
17. Crow TJ: The two-syndrome concept: Origins and current status. Schizophr Bull 11: 471-486, 1985.
18. Andreasen NC: Scale for the Assessment of Negative Symptoms (SANS), Iowa City, IA, University of Iowa, 1984.
19. Beck AT, Beck RW: Screening depressed patients in family practice: A rapid technique. Postgraduate Med 52: 81-85, 1972.
20. Beck AT, Steer RA, Garbin MG: Psychometric properties of the Beck Depression Inventory: Twenty-five years of evaluation. Clinical Psychology Review 8: 77-100, 1988.
21. Andreasen NC, Olsen S: Negative and positive schizophrenia: Definition and validation. Arch Gen Psychiatry 39: 789-794, 1982.
22. Siris SG, Adan F, Cohen M et al: Postpsychotic depression and negative symptoms: An investigation of syndromal overlap. Am J Psychiatry 145: 1532-1537, 1988.

Section 3

SUICIDE AND PROGNOSIS

INTRODUCTION

Bryan Tanney

Suicidal behaviours have increased markedly over the past quarter century in North America and Western Europe. Of special concern is the disporportionate rise in numbers of youthful men (15-29) ending their lives prematurely by suicide. Diverse explanations for this rise in youthful psychological, psychosocial, sociocultural or socio-environmental frameworks. Despite the availability of psychological autopsy studies which clearly suggest an important role of mental disorders in suicide, (1,2) this perspective has been largely disconted. Specifically, the objection is made that these studies have little explanatory power because: 1) they focussed on adult population suicides, 2) they posited a major contribution of affective (mood) disorders and substance abuse, both of which were regarded as disorders of adult life. In the past decade, autopsy evidence from studies of teen and youth suicide (3,4) has renewed interest in the relationships of specific mental disorders to suicide. Psychotic disorders are recognized as important risk factors for suicide in these recent studies (5,6). In many ways, this is not new but rather rediscovered knowledge. Osmond and Hoffer (7) had offered a reminder of the strong link between schizophrenic disorders and suicides that had been noted earlier in the century by Bleuler and others. Miles' review (8) clearly focussed the importance of the schizophrenias as a major at-risk factor for completed suicides. The contributions to this present volume explore the possible role of depression as a mediating variable between schizophrenia and suicide, and leads to widely divergent perspectives on the nature of this linkage. Roy summarizes the relationships between both attempted and completed suicide and schizophrenia, with emphasis on the typical profile of a youthful male early on in their illness. Focusing on the clinical correlates and risk factors in schizophrenics at-risk for suicidal behaviors he acknowledges the presence of depressive symptoms, although not speculating on their nature and/or origins. Instead, the evidence for metabolic (csd SHIAA) or structural (VBR) abnormalities in schizophrenics who have attempted suicide is presented. Roy appears to suggest that specific biological deficits may characterize schizophrenics at high risk of suicide. Although some depressions may exhibit the same abnormal biochemistry, the linkage to depression is not emphasized in these speculations.

In their follow-up of a population of young psychotics, Westermeyer and Harrow report that suicide was equally likely in the schizophrenic and affectively disordered subgroups. A profile of completed suicides as young intelligent and educated leads them to speculate that depression and suicide result from dissatisfaction with the limitations on life and career expectations consequent to schizophrenic illness.

This conclusion is, in fact, the beginning for Drake and colleagues. In previous papers (9,10,11), they have suggested that suicide could be regarded as a realistic response, a "solution" to the depression and "hopelessness" which surface when a patient recognizes the severe limitations of potential which are part of being schizophrenic. In their contribution here, they offer numerous strategies and resources, for dealing both with the schizophrenic illness and with the patient;s responses to it. These proposals could justifiably be viewed as effective suicide prevention strategies in this at-risk population.

Together, these three papers generate an important argument from the sociobiological perspective to explain some of the noted recent increase in young male suicides. The following hypothesis offers one possibility. although there is scant evidence to suggest that the incidence of schizophrenia has changed during this time, there is plentiful indication that our treatment, support and understanding have altered considerably. This is most clearly evidenced in the tragic closure of many publicly supported institutions for chronic mental disorders and their replacement by inadequately funded community mental health programs. The lack of support available to young schizophrenics learning to be with their illness may have directly contributed to their despair and early death.

While offering support for a relationship between the mental disorders of the schizophrenias and suicide, these studies also confound the task of both the psychiatric clinician interested in schizophrenic illness and the suicide prevention resource. The linkage of suicide, depression and schizophrenia is unclear. Several widely divergent explanations can be recognized:

A) Schizophrenia as a chronic debilitating disorder with lifelong limitations of achievement potential for many patients. When not in a period of thought disorder due to the illness, a depressive response to the stress of this illness with features of hopelessness and helplessness would seem realistic. In this view, suicide is a rational response to a quality of life evaluated as no longer worth enduring.

B) Depression as an integral part of the syndrome of schizophrenia, especially in the post-psychotic phase. Treatment of this depressive component is available through biological therapies and through the provision of caring, supportive and insulating environments as the patient

134

gradually re-enters life with a modified and realistic perspective of their limitations.

C) Suicide as a complication of a particular subgroup of schizophrenias. Depression may not be the mediating variable, but may present as a less lethal outcome of the biochemical/neuropsychiatric defect that, in its fullest expression, leads to suicide.

Which view offers a more likely explanation? The issue remains unresolved. Aware of these differing viewpoints, the clinician must struggle with the markedly different intervention each might suggest as a response to a schizophrenic patient contemplating suicide. Is suicide a realistic quality of life decision to be made by a non-psychotic and competent patient? Is it a constricted and temporary view "from the shades" of a depression that is part of the illness itself? Is it a particularly malignant outcome of a subgroup of schizophrenias with identifiable neurological/neurochemical abnormalities?

This conundrum is not only for the caregivers of schizophrenia. It represents part of a larger and ongoing discussion about the nature of suicide and our society's responsibilities to the suicidal person. It reflects our incomplete understanding of the complex and multifactorial origins of the decision for suicide in any single individual. The most likely outcome will validate the heterogeneous nature of suicidal behaviors. It will be accepted that suicide exists for some as a rational response to a hopeless/helpless situation; for others as a disturbed response from the (temporary) perspective of a depression indiced by altered biology. For psychiatric clinicians and all caregivers to those working with schizophrenia and suicide, this conclusion may be an unsatisfactory, incomplete answer when the immediacy of a person at risk of suicide is before them. The papers in this section set forth our present understanding and provide an impressive beginning to much needed further discussion.

REFERENCES

1. Robins E, Murphy GE, Wilkinson RH, Gassner S, Kayes J: Some clinical considerations in the prevention of suicide based on a study of successful suicides. Am J of Public Health 49: 888-889, 1959.
2. Barraclough BM, Bunch J, Nelson B, Sainsbury P: A hundred cases of suicide: clinical aspects. Br J Psychiatry 125: 355, 1974.
3. Shifi M, Carrigan S, Whittinghill JR, Derick, R: Psychological autopsy of completed suicide in children and adolescents. Am J Psychiatry, 142: 1061-1064, 1985.
4. Rich CL, Fowler RL, Young D: San Diego Suicide Study I: Young vs. Old Subjects. Arch Gen Psychiatry 43 (6): 577-582, 1986a.
5. Rich CL, Fowler RL, Young D: Suicide and Psychosis: Changing patterns. Proceedings, American Association of Suicidology, 19th Annual Meeting, Atlanta, GA, p. 94-95, 1986b.

6. Brent DA, Perper JA, Goldstein CE, Kolko DJ, Allan MJ, Allman CJ, Zelenak JP: Risk factors for adolescent suicide. Arch Gen Psychiatry, 45 (6): 581-588, 1988.
7. Osmond H, Hoffer A: Schizophrenia and suicide. J Schizophrenia 1 (1): 54-64, 1967.
8. Miles CP: Conditions predisposing to suicide: a review. J Nerv Ment Dis 164: 231-246, 1977.
9. Drake RE, Gates C, Cotton PG, Whitaker A: Suicide among schizophrenics: who is at risk? J Nerv Ment Dis 176: 613-617, 1984.
10. Cotton PG, Drake RE, Gates C: Critical treatment issues in suicide among schizophrenics. Hosp Commun Psychiatry 36: 534-536, 1985.
11. Drake RE, Cotton PG: Depression, hopelessness, and suicide in chronic schizophrenia. Br J Psychiatry 148: 554-559, 1986.

SUICIDAL BEHAVIOR IN SCHIZOPHRENICS

Alec Roy

National Institute of Mental Health
Bethesda, Maryland

Bleuler (1) described the suicidal drive as "the most serious of schizophrenic symptoms". Also, recent studies suggest that depression is common during the course of schizophrenic illness (2-5). This chapter will examine whether or not there is a relationship between depression and suicide in schizophrenia. Other possible determinants of suicidal behaviour in schizophrenic patients will also be discussed.

SUICIDE

Incidence

Numerous follow-up studies of schizophrenic patients over the last 45 years have reported that schizophrenic illness carries with it an increased risk of suicide. Miles (6) in 1977 reviewed all the then published follow-up studies and concluded that up to 10% of schizophrenics die by suicide. He estimated that at that time in the United States there were approximately 3,800 schizophrenics who committed suicide each year.

There are large variations between studies for the calculated suicide risk for schizophrenics. Using the records of the Houston Veterans Administration Hospital, Pokorny (7) in 1964 calculated that the suicide rate for male schizophrenic patients was 167/100,000/year compared with the then US national rate of about 10/100,000/year. More recently, from the Missouri psychiatry case register, Evenson et al (8) estimated that the age-adjusted suicide rate for male schizophrenic patients was 210/100,000 while the rate for females was lower at 90/100,000. Also in 1982 Wilkinson, using the Camberwell Psychiatric Case Register, estimated that the annual suicide rate for first-admission schizophrenic patients was between 500 and 750/100,000. In reported series of psychiatric patients known to have committed suicide, schizophrenic patients usually account for up to a third of such patients (9-11).

Clinical Correlates

Schizophrenic patients who commit suicide tend to be male and young (Table 1). The first few years of schizophrenic illness are a period of increased risk for suicide. Schizophrenic patients tend to commit suicide in relationship to their last psychiatric hospitalization. Approximately 30 percent of schizophrenic patient suicide victims commit suicide while inpatients, though not usually in the hospital itself. Among schizophrenic outpatients, the first few weeks and months after discharge from a hospitalization are a period of increased suicide risk (Table 2).

Table 1   Age and Sex Ratio in Studies Reported Since 1980 of
           Chronic Schizophrenic Patient Suicide Victims

Author	Number suicides	Mean age (years)	% male
Cheng (36)	12	31.0	83.3
Roy (13)	30	27.9	80
Hogan and Awad (37)	67	35.5	80.6
Breier and Astrachan (38)	20	30.3	90
Drake et al (14)	15	31.7	60
Wilkinson and Bacon (22)	17	42	47
Nyman and Jonsonn (39)	10	30.4	90
Allebeck et al (19)	32	35.9	46.9
TOTAL	203	32	72.2

Depression

Studies reported over the last 30 years strongly suggest that depression is closely associated with suicide in schizophrenic patients. Depressive symptoms have been noted in the last period of psychiatric contact in a substantial percentage of schizophrenic patient suicide victims. There have been nine studies reporting on the presence or absence of associated affective symptoms. Among the total of 270 schizophrenic patient suicide victims in these studies, in approximately 60 percent of them affective symptoms were noted during their period of contact before the patient committed suicide (Table 3). Similarly, in a tenth study, Beisser and Blanchette (12) reported a "high frequency of depression" among 32 schizophrenic patient suicide victims.

Risk factors for suicide

Studies have suggested the likely risk factors for suicide in schizophrenic patients. These include being young

Table 2   Clinical Studies Reporting when During their care
           that Chronic Schizophrenic Patient Suicide Victims
           Committed Suicide

Suicide

Author	Number suicides	As inpatient	After discharge
Cohen et al (40)	40	18 (45%)	22 (55%) on pass, AWOL, trial visit or recent discharge
Pokorny (7)	31	2 (6.5%)	15 (48.4%) within 1 month
Warnes (41)	16	8 (50%)	8 (50%) soon after discharge
Yarden (42)	20	2 (10%)	11 (55%) within 3 months
Roy (13)	30	7 (23.3%)	13 (43.3%) within 6 months
Drake et al (14)	15	5 (33.3%)	7 (46.7%) within 6 months
Wilkinson and Bacon (22)	17	6 (35.3%)	Mean 10.5 months for females and 32 months for males
Allebeck et al (48)	32	14 (44%)	insufficient data
TOTAL	201	62 (30.8%)	

and male, having a relapsing illness, having been depressed
in the past, being currently depressed, having been admitted
in the last period of psychiatric contact with accompanying
depressive symptoms or suicidal ideas, having recently
changed from in- to outpatient care, and being socially
isolated in the community (13).

Recently, Drake and coworkers (14) set out to determine
which of these risk factors distinguished 15 schizophrenic
patient suicides from 89 living schizophrenic patients.
They, too, found that the suicides were young, the majority
were male, had a chronic illness with numerous exacerbations
and remissions (a mean 6.8 admissions during a mean 8.4 years
of illness), and at their last hospitalization significantly
more of them were depressed (80% vs. 48%, P < 0.01), felt
inadequate (80% vs. 36%, P < 0.01), hopeless (60% vs. 27%, P
< 0.01) and had suicidal ideation (73% vs. 38%, P < 0.01).
Seventy percent of the outpatient suicides in their study
killed themselves within 6 months of discharge, and
significantly more of the suicides lived alone (60% vs. 27%,
P < 0.01).   Other studies have also noted that the post-
discharge period is a time of increased suicide risk (13, 14,
42) (Table 4).

Table 3   Clinical Studies Reporting Absence or Presence of
          Depression in Last Period of Contact Among Chronic
          Schizophrenic Patient Suicide Victims

Author	Number suicides	Number and % who were depressed
Levey and Southcombe (43)	23	6 (25.2%)
Cohen et al (40)	40	28 (70%)
Warnes (41)	16	12 (75%)
Yarden (42)	20	13 (65%)
Virkkunen (44)	82	57 (69.5%)
Cheng (36)	12	12 (100%)
Roy (13)	30	16 (53.3%)
Drake et al (14)	15	12 (80%)
Allebeck et al (19)	32	4 (12.5)
TOTAL	270	160 (59.3%)

Table 4   Characteristics of Last Hospitalization Among Twenty
          Schizophrenic Patient Suicides

	N	Percentage
Time elapsed between last discharge and suicide:		
not discharged	2	10
less than 1 week	6	30
1 week < 1 month	2	10
1 month < 3 months	3	15
3 months < 1 year	5	25
Over 1 year	2	10

From Yarden (42)
Published with permission Comprehensive Psychiatry

Prediction of Suicide

These risk factors may well be useful in the acute short term suicide risk assessment of schizophrenic patients. However, unfortunately they are probably of limited value in the long range prediction of eventual suicide. Shaffer and coworkers (15) carried out a five-year follow up study of 361 schizophrenic patients admitted to the Phipps Clinic at Johns-Hopkins Hospital. They found that 12 of these patients eventually committed suicide. However, none of the signs and ratings derived from the case notes of the index admission, either singly or in combinations, accurately predicted the 12 patients known to have committed suicide.

Similarly, we recently carried out a follow-up study of 100 chronic schizophrenic inpatients seen at the National Institute of Mental Health (30). This revealed that 6 of the 100 patients had committed suicide over the mean 4 1/2 years of the follow up period. We used 7 combinations of socio-demographic and clinical variables, recorded at their index NIMH admission, to try to identify these 6 suicide victims. Having had a major depressive episode was one of the variables used. Five of the 7 "strategies" used identified only 1 of the 6 patients who eventually committed suicide; one strategy identified none, and another identified only 2 of the 6 suicide victims.

In our study all the other patients identified by the 7 strategies fell within the group of 94 patients known not to have committed suicide. In fact, 5 of the 7 strategies identified from 3 to 12 patients as at risk for committing suicide who were still alive. Four of the 6 patients who did commit suicide were not identified by any of the 7 strategies used (Table 5).

The difficulty of the long-term prediction of suicide is well recognized. Among psychiatric patients this was recently demonstrated in a large study by Pokorny (17). Twenty-one items were used to identify, among a consecutive series of 4,800 admission, a subsample of 803 patients who were thought to be at high risk for suicide. A five-year follow-up study revealed that 67 of the patients had subsequently committed suicide. Thirty-seven of them were not found among the high risk subsample (false negatives) while 766 of the 803 patients thought to have a high risk for suicide were, in fact, alive (false positives).

In the studies of Shaffer et al and Roy et al, the inability to identify the majority of the schizophrenic patients who subsequently committed suicide supports Pokorny's conclusion that it "...is inescapable that we do not possess any item of information or any combination of items that permits us to identify to a useful degree the particular persons who will commit suicide,..." However, the work of Drake and Cotton (18) has suggested that the depressive symptom of hopelessness is particularly closely associated with schizophrenic suicide victims. The predictive power of this important symptom, either alone or in combination with other variables, has not as yet been further investigated.

Table 5   Seven Combinations of Clinical Features, Determined
           at the Index Admission, used as "Strategies" to try
           to Identify the 6 Schizophrenic patients Determined
           at Follow-up to have Committed Suicide

Category of patient at their index admission	Number of patients identified in each category	Number of patients who subsequently committed suicide
1. Onset under 25 years, past depression, 5 + admissions	10	1
2. Onset under 25 years, past depression, 7 + admissions	5	1
3. Onset under 25 years, past depression, total length of admissions total over a year, 2 or more past suicide attempts	4	1
4. 2 or more past suicide attempts	15	2
5. 3 or more past suicide attempts	6	1
6. 4 or more past suicide attempts	2	1
7. 5 or more past suicide attempts	1	0

From Roy et al (30)
Published with permission Canadian Journal of Psychiatry

Previous Suicide Attempt

     There are eleven studies in the literature reporting the
number of schizophrenic patient suicide victims who had made
a previous attempt (Table 6).   Among the total 289 suicide
victims, 159 (55%) had made a previous attempt.

     There is some evidence that a previous suicide attempt
may be the best indicator that a schizophrenic patient has an
increased risk of committing suicide.   Shaffer et al found
that significantly more of the patients in their follow up
study who eventually committed suicide, than controls, had
made a previous suicide attempt. They also reported that the
number of previous suicide attempts was the single most
important variable associated with eventual suicide.   Also,
in our follow study (30) 5 of the 6 patients who committed
suicide had made a previous suicide attempt by the time of
their index NIMH admission.   Drake et al (14) noted that
significantly more of their 15 suicide victims had made an
explicit suicide threat compared with 89 living schizophrenic
patients (67% vs. 28%, P < 0.01).

Table 6   Studies Reporting Number of Chronic Schizophrenic
           Patient Suicide Victims who had made a Previous
           Suicide Attempt

Author	Number suicides	Number and % who made previous suicide attempt
Cohen et al (40	40	22 (55%)
Warnes (41)	16	6 (37.5%)
Yarden (42)	20	13 (65%)
Shaffer et al (15)	12	5 (42%)
Virkkunen (44)	82	39 (47.6%)
Cheng (36)	12	7 (58.3%)
Roy (13)	30	12 (40%)
Breier and Astrachan (38)	20	12 (60%)
Drake et al (14)	15	11 (73%)
Nyman and Jonsson (39)	10	9 (90%)
Allebeck et al (48)	32	20 (62.5%)
TOTAL	289	156 (54%)

ATTEMPTED SUICIDE

Incidence

        Unfortunately, there is little information about how
many schizophrenic patients attempt suicide. What data there
are come from a variety of different studies. Some studies
report the incidence of suicide attempts among newly admitted
schizophrenic patients, others report the incidence over the
whole course of illness among long-term institutionalized
patients, while others report the incidence among
schizophrenic outpatients over varying follow-up periods
after discharge from an index hospitalization. However,
taken together these various sources of evidence reveal that
a substantial percentage of schizophrenic patients do attempt
suicide at some time during the course of their illness. The
reported percentages vary from 18 to 55.1 percent, depending
on the nature of the study (Table 7).

        Suicidal ideation is also common among schizophrenic
patients. For example, McGlashan (20) followed up 163
chronic schizophrenic patients discharged from Chestnut Lodge
Hospital, in Maryland, between 1950 and 1975. They found
that, over a mean 15 years of follow up, not only had 24
percent of the schizophrenic patients attempted suicide, but
40 percent of them had had suicidal thoughts.

Clinical correlates

        In the general population two to three women attempt
suicide for each man who does. However, among chronic
schizophrenic patients the sex difference does not appear to
be so marked. For example, among the 51 adolescent
schizophrenic patients reported by Inamdar et al (21), 15 of
the 30 boys (50 percent) compared with seven of the 21 girls
(33.3 percent) had a history of suicidal behavior.
Similarly, among 127 adult chronic schizophrenic patients
reported by Roy et al (16), there was no significant
difference between the sexes; 41 of the 73 male patients

143

(56.2 percent) had attempted suicide as had 29 of the 54 female patients (53.7 percent). Surprisingly, in that study there was also no significant difference between the sexes for the methods used in attempting suicide.

In the largest study to date, Wilkinson and Bacon (22) reported on 458 schizophrenic parasuicides seen in Edinburgh between 1968 and 1981. The sex ratio was equal: 230 (50 percent) were male and 228 (50 percent) were female. The male parasuicides were significantly younger than the females (33 versus 37 years) and had received a diagnosis of schizophrenia for a shorter period of time (3.4 versus 7.2 years). Seventy percent of the schizophrenic patients had overdosed with medications prescribed by physicians, and in two-thirds of these patients the drugs had been prescribed for the patient himself. Two thirds of the patients had a previous hospital admission for parasuicide, 91 percent had a prior psychiatric admission, 10 percent had a family history of suicide, and at the time of the suicide attempt 99 of the patients (19.9 percent) had taken alcohol.

Wilkinson and Bacon also matched a random sample of the 124 schizophrenic first-ever parasuicides with a schizophrenic control group. This comparison showed that the male schizophrenic parasuicides had significantly more admissions than male controls, significantly more of them were out of contact with the psychiatric services, and significantly fewer of these schizophrenic parasuicides were receiving neuroleptic medication.

Attempts as inpatients

Some schizophrenic patients attempt suicide while inpatients. In Wilkinson and Bacon's study a quarter of the male, and a third of the female, schizophrenic attempters were either inpatients at the time that they attempted suicide or had been inpatients within the previous month.

Small and Rosenbaum (23) reported on nine psychiatric inpatients who jumped from a height while hospitalized: eight had chronic schizophrenia while the ninth had a schizoaffective disorder. Their average age was 26 years, all were unmarried, socially isolated, and all had a history of five or more previous hospitalizations; five of these nine patients had 10 or more previous hospital admissions. Six of the nine patients had made several previous attempts at suicide. Two thirds had histories of substance abuse, and all but one had been chronically assaultive, especially in the hospital, where all had required frequent restraint and seclusion. All nine patients were receiving neuroleptics. In all nine there had been an apparent precipitant before jumping, such as a change in their treatment plan or social situation, for example, a new therapist or a planned discharge.

Biological factors

There has been a great deal of interest in possible biologic determinants of suicidal behaviour. Asberg et al (24) reported the first study of central monoamine metabolites in patients exhibiting suicidal behaviors. They

Table 7. Number of Schizophrenic Patients who had Attempted
         Suicid Reported in 15 Studies of Living Patients

Author	Number of patients	Type of Study	Number and/or % of patients who attempted suicide
Achte (52)	Unknown	Follow-up	22%
Lehman and Ban (35)	220	Newly admitted	60 (27%)
Cohen et al (40)	40	VA inpatients	22 (55%)
Lehman and Ban (35)	428	Newly admitted	77 (18%)
Planasky and Johnson (45)	205	VA inpatients	52 (25%)
Shaffer et al (15)*	75	Johns Hopkins inpatients	6 (8%)
Inamdar et al (21)	51	Adolescent inpatients	22 (43.1%)
Roy (13)*	30	In first years of illness	11 (36.6%)
Levy et al (43)	32	Male inpatients	12 (37.5%)
McGlashan (20)	163	15 years follow-up	39 (24%)
Roy et al (16)	127	In first 6 years of illness	70 (55.1%)
Drake et al (14)*	89	In first 7.2 years of illness	44 (49%)
Prasad (48)	55	Inpatients	25 (45.5%)
Prasad and Kumar (52)	201	Inpatients	131 (65.2%)
Prasad and Kellner (47)	417	Day patients	193 (46.3%)
	1460	TOTAL	415 (28.4%)

*schizophrenic patients reported as controls in studies of
schizophrenic patient suicide victims

made the observation that significantly more of their depressed patients with "low" CSF levels of the serotonin metabolite 5-hydroxyindoleacetic acid (5-HIAA) had attempted suicide in comparison with those in the "high" CSF 5-HIAA group. This led to the proposal by Asberg et al., that low CSF 5-HIAA levels may be associated with suicidal behavior.

Since then other studies in personality disordered and depressed patients have also reported an association between low levels of CSF 5-HIAA and aggressive and suicidal behaviors, though there have also been negative reports (reviewed in 25). To date there have been four published studies examining CSF levels of 5-HIAA in relation to attempts at suicide among schizophrenic patients (Table 8).

Two of these four studies reported an association between low CSF 5-HIAA levels and suicidal behaviour in schizophrenic patients and two reported no association. However, in the positive report of van Praag et al (26), although the mean postprobenecid CSF levels of 5-HIAA were lower among the patients who attempted suicide, this difference was not statistically significant. In the positive report of Ninan et al (27) the schizophrenic controls were not randomly derived but were chosen to match the suicidal patients for age, sex, and physical characteristics. On the other hand, the limitations of the negative reports of Roy et al (28) and Pickar et al (29) are that the schizophrenic patients had not attempted suicide in the period before their lumbar punctures and few of them had made violent suicide attempts.

Further CSF monoamine metabolite studies are needed comparing schizophrenic patients who have recently attempted suicide with those who have never attempted suicide and controls. Postmortem neurochemical studies of the brains of schizophrenic suicide victims examining levels of serotonin itself, as well as 5-HIAA and the number of serotonergic receptors, are also needed.

Association with violence

It is of note that low CSF 5-HIAA levels have been found to be particularly associated with violent suicide attempts. Low CSF 5-HIAA levels have also been found among violent offenders and arsonists (reviewed in 31). Thus, low CSF levels of 5-HIAA, although an imprecise indicator of diminished central serotonin turnover, are thought to reflect poor impulse control which may manifest itself either as violence towards others or as attempts at suicide. However, there has been a paucity of studies examining for an association between violent and suicidal behaviours among schizophrenic patients.

To date, Inamdar et al (21) have reported the only such study. Three raters reviewed the charts of a consecutive series of 51 schizophrenic adolescents admitted over a 1-year

Table 8. Studies of CSF Levels of 5-HIAA in Relation to
Suicidal Behavior in Schizophrenic Patients

Author	Subjects	Measure of Suicidality	Result
van Praag (26)	10 nondepressed who attempted in response to imperative hal-halucinations, 10 non-suicidal schizophrenics, 10 controls	Recent suicide attempt	Lower CSF 5-HIAA after probenecid in suicide attempters
Ninan et al (27)	8 suicidal, 8 nonsuicidal patients, matched for age and sex	Lifetime history of suicide attempt	Lower 5-HIAA in suicidal patients
Roy et al (28)	26 who had attempted compared with 26 who had not	Lifetime history of suicide attempt	No assoc-iation with 5-HIAA, HVA or MHPG
Pickar et al (29)	13 who had attempted with 15 who had not	Lifetime history of suicide attempt	No assoc-iation with 5-HIAA, HVA or MHPG

period to Bellevue Medical Center. The patients were rated
as violent if all three raters agreed that "they had ever
actually committed serious assault against a person". A
serious suicidal threat, gesture, or attempt was counted as
suicidal behavior. Thirty-four of the 51 patients (66.7
percent) had a history of violence, 22 (43.1 percent) a
history of suicidal behavior, and 14 (27.5 percent) had been
both violent and suicidal. Inamdar et al suggested that the
high occurrence of suicidal and violent behavior was due to
the impact of the schizophrenic illness occurring during the
developmental period of adolescence. However, an alternate
explanation might be that these patients had diminished
central serotonin turnover which manifested itself in both
suicidal and violent behaviour.

Neuropsychiatric factors

Computerized tomography (CT), and other techniques, have
demonstrated that some schizophrenic patients show evidence
of structural brain damage. Recently, Levy et al (32)
reported an association between cerebral ventricular
enlargement on CT scan and attempts at suicide among 32 male

schizophrenic inpatients. Ten of the 12 patients with ventricular (VBRs) brain ratios above 8.4 percent had made a suicide attempt compared with only two of the 20 patients with VBRs below 8.4 percent. Also, the schizophrenic patients with high VBRs had made more violent suicide attempts than the patients who had attempted suicide and who had normal VBRs. Over a 6-month follow-up period there were three violent suicide attempts - two were successful - and all three of these patients had VBRs over 10.8 percent.

Interestingly, Levy et al also reported that the schizophrenic patients with enlarged cerebral ventricles had significantly higher Beck Depression Inventory scores. However, another group (33) was unable to find an association between the VBR and attempts at suicide in schizophrenic patients. Nevertheless, the relationship between cerebral ventricular size, "negative" schizophrenic symptoms, affective symptoms, and suicidal behavior warrants further study. In this regard it is of note that some studies have reported that schizophrenic patients with CT scan evidence of brain atrophy also showed lower CSF levels of 5-HIAA (34).

Instructing hallucinations

Lehmann and Ban (35) were among the first to examine whether schizophrenic patients attempt suicide because of instructing auditory hallucinations or delusions. They examined the records of all the psychiatric patients newly admitted to the Douglas Hospital, Montreal, in the first 6 months of both 1942 and 1962 as to excitement, hallucinations, or delusions occurring simultaneously with suicidal ideation, threats, or attempts. They found that the percentage of schizophrenic patients manifesting suicidal behavior was 27 percent in 1942, before neuroleptics were introduced, and 18 percent in 1962 after neuroleptics were introduced. In neither year did they find a significant correlation between hallucinations and suicidal manifestation.

More recent studies have also suggested that instructing auditory hallucinations are probably not a frequent cause of suicidal behavior in schizophrenic patients. For example, Breier and Astrachan found that none of the 20 schizophrenic patients suicides in their series could be attributed to hallucinated suicidal commands. Also, Drake et al (14) reported that none of their 15 suicides had command hallucinations and Wilkinson and Bacon (22) found no difference between their schizophrenic patient suicide and control groups for a history of auditory hallucinations.

CONCLUSION

Suicidal behaviour is usually multidetermined. However, in some schizophrenics there appears to be an association between depressive symptoms and suicidal behaviour. Suicidal behaviour among schizophrenics mainly occurs as outpatients. This suggests the need for prospective studies of cohorts of schizophrenic outpatients. Such studies might look more closely at predisposing and precipitating factors in the

etiology of depression and suicidal behaviour among
schizophrenics.

In a preliminary such study we compared 18 depressed
schizophrenic outpatients with 18 nondepressed schizophrenic
outpatients (5). All the patients were well controlled on
neuroleptics. As might have been expected, significantly
more of the depressed schizophrenic had attempted suicide at

Table 9. Significant Differences Between Depressed and Non
-depressed Schizophrenic Outpatients

	Depressed schizophrenic N = 18	Non-depressed schizophrenic N = 18	Significance
Number of admissions	6.9	4.1	P < 0.02
Past depressive episode	15	5	P < 0.001
Past treatment for depression	13	5	P < 0.009
Previous suicide attempt	15	7	P < 0.007
Early parental loss	11	5	P < 0.005
Self-esteem	20.9	29.7	P < 0.005
Living alone	11	4	P < 0.02
Number of life events in previous six mos	4.0	1.4	P < 0.0005

From Roy et al (5)

Published with permission British Journal of Psychiatry

some time. Also, significantly more of the depressed
schizophrenics lived alone, had low self-esteem, had early
parental loss, and had had more life events in the six months
before the onset of depression (Table 9). Thus it may be
that depressive symptoms (and suicidal behaviour) in
schizophrenic outpatients, well controlled by neuroleptics,
may occur in those who also have risk factors for depression
and who experience an excess of life events.

REFERENCES

1. Bleuler E: Demential Praecox or the Group of Schizophrenias. New York, International Universities Press, 1950, page 488.
2. Roy A: Depression in chronic paranoid schizophrenia. Br J Psychiatry 137: 138-139, 1980.
3. Roy A: Depression in the course of chronic undifferentiated schizophrenia. Arch Gen Psychiatry 38: 296-300, 1981.
4. Roy A: Depression in chronic schizophrenia. in Psychiatry: The State of the Art. Vol. 1 Edited by Pichot P, Berner P, Wolfe R, et al, Plenum, 1983.
5. Roy A, Thompson R, Kennedy S: Depression in chronic schizophrenia. Br J Psychiatry 142: 465-470, 1983.
6. Miles P: Conditions predisposing to suicide: a review. J Nerv Ment Dis 164: 231-246, 1977.
7. Pokorny A: Suicide rates in various psychiatric disorders. J Nerv Ment Dis 139: 499-506, 1964.
8. Evenson R, Wood J, Nuttall E, et al: Suicide rates among public mental health patients. Acta Psychiatr Scand 66: 254-264, 1982.
9. Roy A: Risk factors for suicide in psychiatric patients. Arch Gen Psychiatry 39: 1089-1095, 1982.
10. Roy A: Suicide and psychiatric patients. Psychiatr Clin North Am 8: 227-241, 1985.
11. Roy A: Suicide in schizophrenia. in Suicide, Edited by Roy A, Baltimore, Williams and Wilkins, 1986.
12. Beisser A, Blanchette J: A study of suicides in a mental hospital. J Dis Nerv Syst 22: 365-369, 1961.
13. Roy A: Suicide in chronic schizophrenia. Br J Psychiatry 141: 171-177, 1982.
14. Drake R, Gates C, Cotton P, et al: Suicide among schizophrenics: who is at risk? J Nerv Ment Dis 172: 613-617, 1984.
15. Shaffer J, Perlin S, Schmidt C, et al: The reduction of suicide in schizophrenia. J Nerv Ment Dis 159: 349-355, 1974.
16. Roy A, Mazonson A, Pickar D: Attempted suicide in chronic schizophrenia. Br J Psychiatry 144: 303-306, 1984.
17. Pokorny A: Prediction of suicide in psychiatric patients. Arch Gen Psychiatry 40: 249-257, 1983.
18. Drake R, Cotton P: Depression, hopelessness, and suicide in chronic schizophrenia. Br J Psychiatry 148: 554-559, 1986.
19. Allebeck P, Varla E, Kristjansson E, Wistedt B: Risk factors for suicide among patients with schizophrenia. Acta Psychiatr Scand 76: 414-419, 1987.
20. McGlashan T: The Chestnut Lodge follow-up study: Part II: Long-term outcome of schizophrenia and the affective disorders. Arch Gen Psychiatry 41: 586-601, 1984.
21. Inamdar S, Lewis D, Siomopolous G, et al: Violent and suicidal behaviour in psychotic adolescents. Am J Psychiatry 139: 932-935, 1982.
22. Wilkinson G, Bacon N: A clinical and epidemiological survey of parasuicide and suicide in Edinburgh schizophrenics. Psychol Med 14: 899-912, 1984.

23. Small G, Rosenbaum J: Nine psychiatric patients who leaped from a height. Can J Psychiatry 29: 139-131, 1984.

24. Asberg M, Traskman L, Thoren P: 5-HIAA in the cerebrospinal fluid. Arch Gen Psychiatry 33: 1193-1197, 1976.

25. Roy A, Agren A, Pickar D, et al: Reduced cerebrospinal fluid concentrations of homovanillic acid and homovanillic acid to 5-hydroxyindoleacetic acid ration in depressed patients: relationship to suicidal behaviour and dexamethasone nonsuppression. Am J Psychiatry 143: 1539-1545, 1986.

26. van Praag H: CSF 5-HIAA and suicide in nondepressed schizophrenics. Lancet 2: 977-978, 1983.

27. Ninan P, van Kammen D, Scheinin M, et al: Cerebrospinal fluid 5-HIAA in suicidal schizophrenia patients. Am J Psychiatry 141: 566-569, 1984.

28. Roy A, Ninan P, Mazonson A, et al: CSF monoamine metabolites in chronic schizophrenic patients who attempt suicide. Psychol Med 15: 335-340, 1985.

29. Pickar D, Roy A, Breier A, et al: Suicide and aggression in schizophrenia: neurobiologic correlates, in Psychobiology of Suicidal Behaviour. Edited by Mann J, Stanley M. Annals of the New York Academy of Sciences, 487: 189-196, 1986.

30. Roy A, Schreiber J, Mozonson A, et al: Suicidal behavior in chronic schizophrenic patients: a follow-up study. Can J Psychiatry 31: 737-740, 1986.

31. Roy A, Virkkunen M, Guthrie S, Linnoila M: Indices of serotonin and glucose metabolism in violent offenders, arsonists, and alcoholics, in Psychobiology of Suicidal Behaviour. Edited by Mann J, Stanley M. Annals of New York Academy of Sciences, 487: 202-220, 1986.

32. Levy A, Kurtz N, Kling A: Association between cerebral ventricular enlargement and suicide attempts in chronic schizophrenia. Am J Psychiatry 141: 438-439, 1984.

33. Boronow J, Pickar D, Nina P, et al: Atrophy limited to the third ventricle in chronic schizophrenic patients. Arch Gen Psychiatry 42: 266-271, 1985.

34. Potkin S, Weinberger D, Linnoila M, et al: Low CSF 5-hycroxyindoleacetic acid in schizophrenic patients with enlarged cerebral ventricles. Am J Psychiatry 140: 21-25, 1983.

35. Lehman H, Ban T: Nature and frequency of the suicide syndrome, twenty years ago and today in the English-speaking community of the Province of Quebec. Laval Med 38: 93-95, 1967.

36. Cheng L: Paper presented at Annual Meeting of Royal College of Physicians and Surgeons, Toronto, Canada.

37. Hogan T, Awad G: Pharmacotherapy and suicide risk in schizophrenia. Can J Psychiatry 28: 277-281, 1983.

38. Breier A, Astrachan B: Characterization of schizophrenic patients who commit suicide. Am J Psychiatry 141: 296-209, 1984.

39. Nyman A, Jonson J: Patterns of self destructive behaviour in schizophrenia. Acta Psychiatr Scand 72: 252-262, 1986.

40. Cohen S, Leonard C, Farberow N, et al: Tranquilizers and suicide in the schizophrenic patients. Arch Gen Psychiatry 11: 312-321, 1964.

41. Warnes H: Suicide in schizophrenia. _Dis Nerv Syst_ 29, 35-40, 1986.
42. Yarden P: Observations on suicide in chronic schizophrenics. _Compr Psychiatry_ 15: 325-333, 1974.
43. Levy S, Southcombe R: Suicide in a state hospital for the mentally ill. _J Nerv Ment Dis_ 117: 504-514, 1953.
44. Virkkunen M: Suicide in schizophrenia and paranoid psychosis. _Acta Psychiatr Scand (Suppl)_ 250: 1-305, 1974.
45. Planasky K, Johnson R: The occurrence and characteristics of suicidal preoccupations and acts in schizophrenia. _Acta Psychiatr Scand_ 47: 473-483, 1971.
46. Prasad A: Attempted suicide in hospitalized schizophrenics. _Acta Psychiatr Scand_ 74: 41-42, 1986.
47. Prasad A, Kellner P: Suicidal behaviour in schizophrenic day patients. _Acta Psychiatr Scand_ 77: 488-490, 1988.
48. Allebeck P, Varla A, Wistedt B: Suicide and violent death among patients with schizophrenia. _Acta Psychiatr Scand_ 74: 43049, 1986.
49. Drake R, Gates C, Whitaker A, Cotton P: Suicide among schizophrenics: a review. _Comp Psychiatry_ 26: 90-100, 1985.
50. Drake R, Gates C, Cotton P: Suicide among schizophrenics: a comparison of attempters and completed suicides. _Br J Psychiatry_ 149: 784-787, 1986.
51. Cotton P, Drake R, Gates C: Critical treatment issues in suicide among schizophrenics. _Commun Psychiatry_ 36: 534-536, 1985.
52. Prasad A, Kumar R: _Suicide Life Threatening Behaviour_, 1988, in press.

EARLY PHASES OF SCHIZOPHRENIA AND DEPRESSION:

PREDICTION OF SUICIDE

Jerry F. Westermeyer and Martin Harrow

Michael Reese Hospital and Medical Centre
Departments of Psychiatry and Behavioral Sciences
University of Chicago, Chicago Illinois

> "The Church knows all the rules. But
> it does not know what goes on in a
> single human heart."
>
> Graham Greene The Heart of the Matter

As prospective studies of representative samples of schizophrenics and other types of psychotic and nonpsychotic patients have increased (1-5), suicide has been recognized as a major problem which merits greater concern on the part of clinicians and policy makers alike. For mentally disturbed samples, literature reviews suggest that the suicide rate may range from 10% to 15% among the major mental disorders (1,4-5). For the general population, suicide is a leading cause of death among adults under forty years of age, and suicide rates have been increasing since 1955 with a slight decrease in the late 1970s (6).

However, with some notable exceptions (3,7-10), suicides have not been reported in a number of major outcome studies of schizophrenia, and suicide rates have not been documented in some reviews of schizophrenic outcome (11,12). Furthermore, there have been relatively few well-controlled prospective studies of schizophrenia in which risk and protective factors for suicide can be explored (1). Such studies may be of great help to clinicians in identifying patients at risk for suicide and making necessary treatment interventions. One possibility is that risk factors for suicide may vary according to diagnosis (1,4,13). Needed are prospective studies in which the prognostic factors of patients who suicide across a range of different diagnostic groups can be compared to each other as well as to control groups of patients who do not commit suicide.

The present research explores rates of suicide in schizophrenia and other major diagnostic disorders such as depression and examines several identifying characteristics of persons at risk for suicide. It improves upon many previous studies on the prediction of suicide by employing a prospective research design in which patients were assessed at index hospitalization before any of them had committed suicide.

In addition, young patients in early phases of their illness were chosen as subjects in this research to study the full course of their disorders across the adult life cycle. This may have important implications for the findings of this report because suicide rates for different diagnostic disorders may vary across the adult life cycle. The following questions were addressed in this chapter:

1) What is the suicide rate among young, early phase DSM III schizophrenics?

2) Is the suicide rate high among other types of psychotic patients (e.g., psychotic depressives) and nonpsychotic patients (e.g., nonpsychotic depressives)?

3) What types of schizophrenic patients are at risk to commit suicide?

4) Is there a critical period during the follow-up when suicide is likely to occur?

METHOD

The current research is based on longitudinal data from two samples of patients. These include a sample of 400 patients from the Chicago Follow-up Study, and a sample of 186 patients studied earlier as inpatients (14) and followed-up by Grinker, Harrow and Westermeyer. Since the data which emerged from these two samples were similar, and were based on the same type of data collection, they were combined in this report. The latter of these two follow-up studies, The Chicago Follow-up Study, was designed to analyze:

1) Prognosis and outcome over time in a range of psychiatric patients, but focusing primarily on the course of adjustment of patients with schizophrenia (7, 15-20); and

2) Possible mechanisms that may be involved in thought pathology and psychosis (21-15).

Patient Sample

The total sample included 586 young psychiatric patients studied initially at index hospitalization, prior to subsequent follow-ups to assess posthospital adjustment and potential suicides. The patients in both samples were assigned DSM III and RDC diagnoses, utilizing an extensive inpatient data collection system. These diagnoses were based on a detailed admission interview summarized in the patient's chart, on the Schedule for Affective Disorder and Schizophrenia (SADS) (26), and on the Schizophrenia State Interview (SSI), a semi-structured, tape-recorded interview (14). Using DSM III diagnostic criteria, the samples include

135 schizophrenics, 194 other psychotic disorders and 257 nonpsychotic disorders.

The patients averaged 23 years of age and 76% of this young sample had never married at index hospitalization. It was originally hoped that these patients would be followed-up across the adult life cycle and consequently young patients in early phases of their disorder were chosen for study. Sixty percent were from the higher parental social classes (i.e., I, II or III) using the Hollingshead-Redlich system (27). The patients averaged 13 years of education at hospital admission. Fifty-one percent of the total sample were first admission patients and most of the remainder of the sample had only experienced one or two hospitalizations prior to index hospitalization. In general these young, relatively early phase patients had not experienced a chronic course of numerous psychiatric hospitalization extended over many years prior to entry into the current study.

Fifty-two percent of the total sample were females. Males dominated the schizophrenic disorders (i.e., 64%) and females dominated the nonpsychotic disorders (i.e., 61%) which were primarily comprised of nonpsychotic depressive disorders.

All patients were prospectively assessed at index hospitalization and diagnosis and prognostic factors were rated blindly as to subsequent outcome. Major predictors included items assessed in relation to risk for suicide, as well as diagnostic subtype (e.g., paranoid vs. nonparanoid), demographic items (e.g., sex, age, social class, marital status, race and education), the Phillips Scale of Premorbid Adjustment (16) and the Vaillant-Stephens Classical Prognostic Scale (12,18,28). The classical prognostic items included the assessment of 1) affective symptoms at hospitalization (e.g., depressive mood, guilt, death ideation or concerns), 2) longitudinal factors assessing duration of symptoms and functioning prior to hospitalization (e.g., acute onset, work and social history) and 3) other types of information such as intelligence. Estimates off IQ or intellectual performance were obtained by using the Information Subtest of the WAIS (29) which was given during index hospitalization to the 400 patients from the Chicago Follow-up Study.

Attempts were made to contact all patients for follow-up interviews during the posthospital period after hospitalization. Ten percent of the total sample could not be located for follow-ups in the posthospital period. The major estimate of rate of suicide which we used was based on how many patients had clear evidence of suicide from among the entire sample of patients originally assessed at index hospitalization. Since there could have been a few additional patients who committed suicide among patients that could not be located for follow-up, the suicide rate we obtained may be a slight underestimation.

Forty-seven patients died in the posthospital period and 33 of these 47 patients (or 70% of patients who died in the early posthospital period) committed suicide. Suicide was established by either: 1) death certificate or 2) the presence of both a) and b) as follows: a) family report of suicide __and__ b) method of death which made suicide extremely likely (e.g., falling from 23rd floor of building after expressing death wishes). The 33 patients who had suicided constituted 5.6% of the total sample.

RESULTS AND DISCUSSION

A. The Risk for Suicide in Early Phase, Young
    Schizophrenics

    The sample of 33 patients who committed suicide were compared to the patient groups which did not commit suicide. Table 1 compares the diagnostic groups on rate of suicide.

Table 1   Prediction of Suicide Among Young Psychiatric
           Patients by DSM III Diagnosis

DSM III DIAGNOSES	PATIENTS WHO COMMITTED SUICIDE
SCHIZOPHRENICS	13 (9.6%)
OTHER DSM III PSYCHOTIC DISORDERS	12 (6.2%)
NON-PSYCHOTIC DSM III DISORDERS	8 (3.1%)

$x^2 = 7.24$, (2 df), $\underline{p} < .05$

    The results show that thirteen DSM III schizophrenics (or 9.6% of the total sample of 135 schizophrenics initially assessed at hospitalization) committed suicide. Eleven (or 85%) of the 13 schizophrenics who suicided were men. The following case history illustrates a schizophrenic suicide:

> Mr. X, a 27 year old, white, single schizophrenic man was originally hospitalized with paranoid ideation and hallucinations at index hospitalization. He reported: "I was scared of my parents and I was scared of the FBI. I was scared of being shot." Mr. X's extensive paranoid delusions seem to have troubled him for some time. He described his childhood as "lonely with no close friends". He said he handled hallucinations by "isolating" himself thus avoiding "overstimulation". He graduated from high school and worked as a photocopier for his father. There was no evidence of suicide attempts or ideation prior to admission. However, he reported being very unhappy in his life. He also reported very low self esteem and being angry at himself. Mr. X suicided at age 30 three years after hospital discharge by shooting himself in the head in his home.

B.   The Risk for Suicide in Other Types of Psychotic and Nonpsychotic Patients and Diagnostic Comparisons

Twelve (or 6.2%) of the 194 other psychotic disorders assessed at index hospitalization committed suicide in the posthospital period.   These results showing a high rate of suicide among other psychotic patients support earlier findings by Tsuang and colleagues (3).   The other psychotic patients who committed suicide included eight psychotic major depressive disorders, two schizoaffective disorders, one psychotic manic disorder, and one schizophreniform disorder. Eight (or 75%) of the 12 other psychotic patients who committed suicide were men.

The schizophrenic group was not more likely to commit suicide than the other psychotic group.   Many patients in the other   psychotic   group   were   formerly   categorized   as schizophrenic using broader DSM II concepts of schizophrenia (18).

The following case history illustrates a suicide from among the other psychotic disorders:

> Mr. Y was a 22 year old, white, single, Catholic man in his junior year of college when he was first hospitalized with a major depressive disorder, psychotic type.   Mr. Y dates the onset of his illness 1 1/2 years ago when he became "paranoid" following marijuana use.   Two months prior to admission he flunked out of college and he experienced a "deep depression" and he "couldn't communicate with anyone".   At hospital admission he showed evidence of hallucinations and psychotic ideation centered around depressive themes, self reproach, depression, extreme fatigue, and inability to concentrate.   Mr. Y described episodic suicide ideation with concrete plans to kill himself with a "gun or in my car".   However, at hospitalization he had not made a suicide attempt because he felt it was not worth the effort.   Mr. Y said "If you are normal you try things, there must be a will to do it (suicide) and I don't have it." Despite this, Mr. Y committed suicide about a year following hospital discharge, at age 23, by shooting himself in the head.

Eight (or 3.1%) of the 257 nonpsychotic disorders assessed at index hospitalization committed suicide in the posthospital period.   This group included 6 nonpsychotic major depressive disorders, one anorectic disorder and one borderline disorder.   Two (or 25%) of the 8 nonpsychotic patients who committed suicide were men.

The following case history illustrates a suicide among the nonpsychotic disorders:

> Ms. Z was a 26 year old, white, single woman who was first hospitalized with a nonpsychotic major depression.   She reported being depressed for a year following a breakup with her boyfriend.   After

a failed attempt to reconcile with this boyfriend, Ms. Z decided she did not want to live and attempted suicide by a medication overdose. She was comatose for two days and was hospitalized. At hospitalization she revealed extreme depression, low self esteem, extreme anxiety, sleep problems and difficulty concentrating. A one year follow-up found her rehospitalized and severely depressed. During the follow-up year, she had been working full time, was dating regularly and had two close friends. However, she was extremely depressed, felt wretched and had made concrete plans to kill herself. She committed suicide one month after the follow-up interview at age 27 of a sleeping pill overdose in her home a few weeks following hospital discharge.

Important diagnostic differences were evident in the rate and type of suicide. Nonpsychotic patients were significantly less likely to suicide than the schizophrenic group (Chi square = 7.33, p < .01). This diagnostic difference held when males from the schizophrenic and nonpsychotic groups were analyzed separately and compared to each other. The data comparing schizophrenics with nonpsychotic patients indicate that for young, early phase psychiatric patients, the presence of schizophrenia is a major risk factor for suicide, particularly among male schizophrenics.

C. Suicide for Psychotic Major Depressives Versus Nonpsychotic Major Depressives

The group of nonpsychotic patients did not show a significantly different rate of suicide in comparison to the other psychotic patients. However, when we focus exclusively on the depressive disorders, the nonpsychotic depressive patients were less likely to commit suicide than the psychotic depressive disorders (Chi square = 4.99, p < .05). Table 2 compares the suicide rate for the major depressive disorder, psychotic type with the nonpsychotic major depressive disorders. Eight (or 12%) of the 69 patients with psychotic major depression committed suicide. In contrast, 6 (or 4.1%) of the 145 nonpsychotic major depressives committed suicide in the posthospital period. The case histories of Mr. Y and Ms. Z presented earlier illustrates a psychotic depressive who suicided (Mr. Y) and a nonpsychotic depressive who suicided (Ms. Z).

D. Method of Committing Suicide Among Diagnostic Groups

Data on the method of committing suicide, used by male and female patients within each diagnostic group, are presented in Table 3. The type of method is categorized according to high and lower lethality employing a modification of a system used by Breier and Astrachan (30), based on categories defined by Tuchman and Youngman (31).

Table 2   Prediction of Suicide Among Psychotic and
           Nonpsychotic Major Depressives

TYPE OF DEPRESSIVE DISORDERS	PATIENTS WHO COMMITTED SUICIDE
MAJOR DEPRESSIVES (PSYCHOTIC)	8 (12%)
MAJOR DEPRESSIVES (NONPSYCHOTICS)	6 (4.1%)

$x^2 = 4.99$, (1 df), $p < .05$

The data on method of suicide indicate that schizophrenic men were more likely to employ highly lethal

methods than the combined group of psychotic and nonpsychotic nonschizophrenic men ($p < .05$). The most common methods for suicide among the schizophrenic group were hanging (n = 5) and shooting oneself (n = 3).

E.   Critical Period For Suicide

Another important factor we explored in relation to risk for suicide was the timing of its occurrence in the posthospital period. Suicide for the schizophrenics was more likely to occur within the first four years following index hospitalization. In contrast, the last follow-up contact for the nonsuicides was most likely to occur five years or more since hospitalization (Chi square = 17.8, $p < .01$). In accord with the results from other studies of suicide among the schizophrenics (1, 4), suicide among schizophrenics is beginning to diminish as the follow-up period lengthens. This trend held for the combined group of other psychotic patients and schizophrenics. Thus, 21 of the 25 psychotic patients (or 84%) who suicided did so within the first four years following index hospitalization. In contrast, suicide among the nonpsychotic patients was not usually confined to the first four years following hospital discharge and may increase as the follow-up period lengthens. Five of the 8 nonpsychotic patients (or 63%) who committed suicide did so five or more years after hospital discharge.

We do not expect that the high suicide rate thus far observed for the young psychotic patients in this sample will continue at the same high rate for the remainder of their life cycle. However, the fact that psychotic patients have not suicided within the first four years of hospital discharge does not make them immune to suicide in later phases of their disorder. Only two suicides among the schizophrenic group occurred many years after hospital discharge. The following case history, involving a patient whose suicide we recently learned about, is one of these later occurring suicides:

Table 3 Type of Death by Diagnostic Group Among Suicides

TYPE OF DEATH	SCHIZOPHRENICS		PSYCHOTICS		NONPSYCHOTICS	
	MALES	FEMALES	MALES	FEMALES	MALES	FEMALES
A. HIGH LETHALITY						
HANGING	5	0	0	0	0	0
SHOOTING ONESELF	2	1	1	0	0	1
DROWNING	2	0	0	0	0	0
SELF-IMMOLATION	1	0	0	0	0	0
JUMPING IN FRONT OF TRAIN	0	0	2	0	0	0
JUMPING FROM GREAT HEIGHT	0	0	1	2	0	2
TOTAL - HIGH LETHALITY	10	1	4	2	0	3
B. LOWER LETHALITY						
DRUG OVERDOSE	1	0	2	1	1	1
INHALING CARBON MONOXIDE	0	0	1	0	0	1
CUTTING SELF	0	0	0	0	1	0
TOTAL - LOWER LETHALITY	1	0	3	1	2	2
C. OTHER TYPE OF DEATH OR UNKNOWN	0	1	1	1	0	1
D. TOTAL SAMPLE	11	2	8	4	2	6

* Significant difference between schizophrenic men and nonschizophrenic men (i.e., both psychotic and nonpsychotic) in type of death (i.e., high lethality vs. lower lethality) (p < .05).

Mr. Q was a 29 year old, white married schizophrenic male who had showed adequate social adjustment prior to his first schizophrenic break. He was hospitalized with delusions of grandeur, paranoia and auditory hallucinations. He was preoccupied with thoughts of self mutilation as a way of absolving himself. He had made several gestures of suicide in the past by cutting himself. Three follow-ups over 10 years found him disorder. He lost several jobs but kept trying to find work and trying fo function well. Supported by his therapist, he did succeed in finding work several times and in a few different instances was successfully employed for years. He owned his own car and home. At the third follow-up he was still married with two sons. Although he claimed that he had close friends, he was socially inactive and spent most of his time alone in his room smoking and listening to music. He suffered extreme anxiety, depression and guilt. He reported four suicide attempts in the posthospital period noting that the last one was "just to show people I am hurting, nothing serious". Mr. Q committed suicide at age 41, twelve years after index hospitalization, by hanging himself in his home.

The high suicide rate for the schizophrenic and other psychotic groups can only rise over time, although hopefully at a much lower rate. Some of the surviving subjects have made suicide attempts in the posthospital period. Other aspects of our data suggest that those schizophrenics at greater risk for suicide may be the more intelligent, striving individuals who originally had higher expectations for their future and who maintain a capacity for depression (such as Mr. Q presented above).

F. Risk Factors for Suicide Among Schizophrenics

To analyze risk factors for suicide among schizophrenics, the 13 schizophrenics who committed suicide were compared to the sample of schizophrenics who did not commit suicide in the posthospital period. The major results are presented in Table 4. The size of the groups vary for several of the risk factors because a number of prognostic items were not rated for all schizophrenics. For example, we were not able to obtain formal test data for all subjects for intellectual ability at hospitalization. Prognostic items which we assessed included a number of major subtypes of schizophrenia, such as the paranoid-nonparanoid subtypes, sex, race, educational level and others.

The overall results indicated that paranoid subtype and race (white) significantly predicted completed suicide among the schizophrenic group. Ten paranoid schizophrenics committed suicide and paranoid schizophrenics were more likely than nonparanoid schizophrenics to commit suicide ($p < .05$). All 13 of the schizophrenics who committed suicide were white ($p < .05$). Race is also a significant predictor of suicide in the general population as whites are more likely to commit suicide than blacks (6). In addition, among these relatively early phase, young schizophrenics, those with a gradual onset of illness were more likely to commit suicide ($p < .05$).

Table 4   Prediction of Suicide Among Schizophrenics

MAJOR TYPES OF SCHIZOPHRENIA	SUICIDE	NONSUICIDE	SIGNIFICANCE
1) Paranoid Subtype			
Paranoid:	15%	85%	
Nonparanoid:	4%	96%	$p < .05$
2) Sex			
Male:	12%	88%	
Female:	4%	96%	$p < .10$
3) Marital Status			
Ever Married:	4%	96%	
Never Married:	11%	89%	N.S.
4) Parental Social Class			
I-III:	16%	86%	
IV-V:	7%	93%	N.S.
5) Race			
White:	14%	86%	
Black:	0%	100%	$p < .05$
6) First Admissions			
First:	6%	94%	
One or More:	15%	85%	N.S.

	SUICIDE Mn (SD)	NONSUICIDE Mn (SD)	SIGNIFICANCE
7) Process-Reactive: (Phillips Scale of Social Competence)	15 (4.9)	15.1 (6.5)	N.S.
8) Number of Previous Hospitalizations:	2.7(2.0)	1.8 (2.1)	N.S.
9) Intellectual Performance WAIS Information Scale:	19.2(6.8)	15.4 (5.4)	N.S.
10) Acute Onset:[+]	3.5(.7)	3.1 (1.0)	$p < .05$

[+] Low score represents a more acute onset.

As noted earlier, male schizophrenics were more likely than female schizophrenics to commit suicide (i.e., 11 of the 13 schizophrenic suicides were male), although this trend was not statistically significant ($p < .10$). The majority (i.e., 64%) of our DSM III schizophrenics were male. Recent research reports have shown that DSM III samples of schizophrenics (those defined according to narrow concepts of schizophrenia) are more likely than DSM II samples of schizophrenics (those defined according to broad concepts of schizophrenia) to be comprised of a larger percentage of males than females (18, 32).

The results indicating a higher rate of suicide for males have been found in several other studies of suicide in schizophrenia (30, 33, 34). Males are also more likely than females to commit suicide in the general population (Hollinger, 1987). In the current study, schizophrenic males often chose efficient, highly lethal ways to commit suicide (See Table 4).

In addition to sex, the results indicate that a number of other suicide risk factors among schizophrenics that have been reported in the literature were not significantly predictive of suicide in the current study. These items include age (younger) and marital status (unmarried). However, age restrictions may have made prediction of suicide more difficult for this group of schizophrenics. Since age was controlled in the present study and few schizophrenics in the total sample had married, these factors failed to be significantly associated with subsequent suicide, even though the average age at the time of suicide among the schizophrenics was 27 years, and 12 of the 13 schizophrenics who committed suicide had never married.

The following case histories illustrate three schizophrenic patients in our sample who committed suicide:

> Mr. J was a 19 year old white, single man with a very poor prehospital social and work history. When he was hospitalized following a suicide attempt, he had auditory hallucinations and delusions. He achieved very high scores on intelligence testing at hospitalization. Mr. J said: "I am here because I am sick". He stated that he has been emotionally disturbed all his life. He was very paranoid and claimed he had been taken to outer space. Mr. J stated that he had a "futile existence" and that he felt dead. He stated that he might as well be dead because he hated the way his life was going, and there did not seem to be any hope for change. Mr. J committed suicide while in the hospital by hanging himself at age 19.

> Ms. I was an 18 year old, white, single woman when she was hospitalized with auditory hallucinations, thought disorder, paranoid ideas, and sleep disturbance. She had not made any previous suicide attempts. She dated the onset of her illness to two years prior to hospitalization following a move from another city. At a two year follow-up, Ms. I reported two suicide attempts during the past year by slashing her wrists and an overdose of medications. During the follow-up period she had dropped out of school and had worked less than half the time, but she was socializing with friends and dated more than once a week. At follow-up, Ms. I was depressed and anxious but she was not delusional or hallucinating. Ms. I committed suicide at age 22, four years after hospital

discharge, by shooting herself in the head in her
home.

> Mr. H. was an 18 year old white, single man with
> schizophrenia when he was hospitalized in the fall
> of his first year away from home at college. He
> experienced both paranoid delusions and a variety
> of hallucinations for 10 months prior to hospital
> admission. He had considered killing himself on
> two occasions before admission and went so far as
> to load a gun. He said he had felt hopeless since
> his father's death one year prior to admission
> because his father was the only one he could look
> up to. He felt "my future was shattered" with his
> father's death, and his suicide would "get rid of
> my agony". Mr. H committed suicide at age 21,
> three years after hospital discharge, by shooting
> himself in the chest.

A major problem in studying completed suicide is that it
may be difficult to know the person's precise motivations and
state of mind at the time of suicide. The person who commits
suicide takes his life story with him or her. However, based
on data collected at index hospitalization and from our
follow-ups, it appears that suicide is often associated with
failed expectations, intense unhappiness by patients about
their current life circumstances and despair about their
future prospects. For example, Mr. J and Mr. H above
expressed deep dissatisfaction about their lives and a
profound sense of hopelessness for the future. In accord
with other research reports (1,5,7), suicide in schizophrenia
often appears to be closely associated with a reaction of
unhappiness, depression and even bitterness about a
debilitating mental illness rather than with delusional or
psychotic behavior. Suicide often seemed to be a calculated
decision by which troubled persons see themselves as being
able to escape situations they view as intolerable.

G. Risk Factors for Suicide Among the Combined Sample of
   Schizophrenics and Other Psychotic Patients

To identify characteristics of suicide for psychotic
patients in general, the combined sample of schizophrenics
and other psychotic patients who committed suicide were
compared on the prognostic and risk factors to the combined
sample of schizophrenics and other psychotic patients who did
not commit suicide. Most of the variables analyzed for the
other psychotic disorders showed the same general trends
(i.e., the same direction) as the variables analyzed for the
schizophrenics. The combined sample of 25 schizophrenic and
other psychotic patients who suicide were significantly more
likely to be men, to be white, to be single, to have higher
IQ scores, and to have a gradual onset of illness.

The following case histories illustrate suicide among
the other psychotic patients who were not schizophrenic, when
viewed in terms of modern, narrow, diagnostic criteria for
schizophrenia:

> Mr. C was an 18 year old, single, white man with an
> excellent social and work history when hospitalized

with a psychotic depressive disorder. Following a lay off from work for seasonal reasons one month prior to admission, Mr. C experienced mood congruent delusions, hallucinations, guilt and sleep and eating disturbances. He made three suicide attempts during that month. Mr. C stated: "I feel like I am in a deep dark hole, nothing I do is right since I was laid off. I don't like the things I used to do. I don't deserve my girlfriend. I don't deserve to live". Mr. C committed suicide 2 1/2 years after hospitalization, at age 21, by inhaling auto exhaust.

Ms. A was a 21 year old white single woman with a history of cutting herself when hospitalized with a psychotic depression. She experienced various types of depressive symptoms and felt there was a conspiracy against her. She was very anxious and feared "falling apart". Ms. A committed suicide at age 22, one year after hospital discharge and two years after her mother's death, by a drug overdose. She left the following note to her therapist: "Dear ...., I have just ingested one-half gallon of wine and all my medications plus 72 sleeping pills. I told you last night my mother is calling me and I just have to answer....By the time you read this I hope I am dead. You see this is one time when I am serious or I would have called you by now. I have never felt this alone in my life.....It was my own decision and I made it. The urge is just too strong."

An examination of case histories of all patients who suicided suggests much diversity in the timing, personal motivations and circumstances by which individuals take their own lives. To a great extent, each life story is unique and generalizations regarding suicide are difficult. However, as with many schizophrenic case histories, a number of the case histories of other psychotic patients indicated failed expectations and a deep dissatisfaction with current life circumstances. As the self reports of Mr. C and Ms. A above indicate, suicide was often viewed as a way out of intolerable or painful circumstances.

DISCUSSION

The primary finding of this research was the alarmingly high rate of suicide among young, relatively early phase patients with schizophrenia and other psychotic disorders, with the schizophrenic group having a significantly greater suicide rate than the non psychotic group. The magnitude of suicide in schizophrenia and other psychotic disorders (especially psychotic depressives) may have been obscured in follow-up research which focused on retrospectively studied samples. In addition, research based on schizophrenics who have been chronically impaired for many years and are older may have underestimated the rate of suicide in schizophrenia. As suicide among schizophrenics is more likely to occur in early phases of the illness, estimates of the total rate of suicide in schizophrenia would be distorted when they are based on older, chronically impaired schizophrenics, and when young, relatively early phase schizophrenics who have already suicided are not included in the sample studied.

The high suicide rate we found among schizophrenic patients is in accord with other research reports (1,2-5) and with our earlier evidence based on a smaller subsample of schizophrenic patients (7). Similarly, the higher than expected rate of suicide in other psychotic patients is in accord with previous findings reported by Tsuang (3,9). Moreover, males appear to be at considerable risk for suicide, both within schizophrenia and with the other psychotic disorders in general.

Young, white, single, male schizophrenics, of paranoid subtype and with gradual onset, seem to be especially vulnerable to suicide in early phases of their illness. In addition, schizophrenics with higher IQ and higher education seem to be at greater risk for suicide. As discussed by Robert Drake (1), Alec Roy (5), and our own research group (7), such patients seem to have high expectations for their lives which are severely damaged by the realities of psychotic illnesses, and their suicide is closely associated with depression.

The association of high IQ and white race with suicide among psychotic patients supports the concept of high expectations as a major risk factor for suicide among schizophrenics. In addition, several follow-up studies have shown that life adjustment is very poor for most schizophrenics (3,7,8,10,20). Schizophrenics are more likely than other types of major mental disorders to experience severe symptomatology, unemployment, and a range of financial and social deprivations (3,7,8,10,20). Some schizophrenics may experience a profound sense of hopelessness and despair as a result of these difficult circumstances, and some of these schizophrenics may be willing to act on their despair. This despair may lead to successful suicide attempts - especially for young, white, often intelligent men who initially had high hopes for themselves.

Moreover, paranoid schizophrenics may be more likely than nonparanoid schizophrenics to be rigid about their original expectations, to clearly perceive their difficult prospects for the future, and to project their fears and feelings of despair on the world in general. The paranoid schizophrenic may be uncompromising or inflexible about potential options for their future. Thus, they may view suicide as their only escape from a difficult or painful situation. The paranoid schizophrenics in this study appeared very determined in ending their lives, and they often chose highly lethal methods for their suicide. Following the theoretical dynamics of paranoia, they occasionally ended their lives in the very ways they had once feared (or projected) others would murder them. (See the case history of Mr. X, above).

The high rate of schizophrenic suicides thus far observed in young adulthood may decline in mid-life and older adulthood. The older, chronic patients probably adjust to their difficult circumstances over the years. The chronic patients may have revised and lowered their expectations for their future. Such patients may be less likely to commit suicide because they do not experience the same sense of abject despair or intense dissatisfaction about their lives as do patients in early phases of their disorder. Thus, a critical period for suicide in schizophrenia appears to be in the first few years of severe illness when the individual must come to grips with what may be viewed as limited prospects for career success or other types of achievements in life.

The diagnostic differences in suicide thus far observed may disappear over time. As other research reports (and our own data) have suggested, suicide among nonpsychotic depressives may increase on a cumulative basis with age. While suicide across diagnostic groups may share some common factors, the rates, timing and select characteristics may vary to some extent for different diagnostic groups. Thus, we would propose that while there are many risk factors for suicide which cut across various types of disturbed people and various diagnostic groups, there are also some risk factors which are diagnosis specific, or specific to certain subgroups such as psychotic patients. In addition, the final suicide rates for this cohort await later follow-ups, and this information could show a different picture.

Finally, it should be emphasized that the study of suicide is quite complex and many different factors may enter into an individuals decision to commit suicide. Moreover, as Pokorny and other researchers have indicated (13,35), it is very difficult to accurately predict specific subjects at increased risk to commit suicide because suicide occurs in such a minority of patients. Consequently, any predictor associated with suicide also identifies many false positives, that is subjects with a risk factor who will not commit suicide. For example, although a high percentage of young, intelligent, white, male schizophrenics will commit suicide, most will not.

Despite major difficulties in predicting suicide, a general description of which types of patients are at increased risk for suicide, in association with clinical judgement and understanding of the individual, may help guide the treatment of vulnerable patients. The current data and increasing data by others (1,3-5) suggest that young schizophrenic patients, and probably other types of psychotic patients in general, are especially vulnerable to suicide. The high vulnerability to suicide among psychotic patients is particularly prominent for better educated, intelligent, white males <u>in the earliest phases of their illness</u>, who initially had high expectations for themselves, and are profoundly dissatisfied with the quality of their lives.

ACKNOWLEDGMENT

This research was supported, in part, by Grant No. 26341 from the National Institute of Mental Health and research grants from the John D. and Catherine T. MacArthur Foundation and the Four Winds Research Fund.

Farzin Yazdanfar provided important computer and statistical assistance in the preparation of this chapter.

REFERENCES

1. Drake R, Gates C, Whitaker A, Cotton P: Suicide among schizophrenics: a review. Comp Psychiatry 26: 90-100, 1985.
2. Fawcett J, Scheftner W, Clark D, Hedeker D, Gibbons R, Coryell W: Clinical predictors of suicide in patients with major affective disorders: A controlled prospective study. Am J Psychiatry 144: 35-40, 1987.
3. Tsuang M: Suicide in schizophrenics, manics, depressives and surgical controls: a comparison with the general population suicide mortality. Arch Gen Psychiatry 35: 153-155, 1978.
4. Miles CP: Conditions predisposing to suicide: a review. J Nerv Ment Dis 164: 231-246, 1977.
5. Roy A: Suicide in schizophrenia, in Alec Roy (Ed.), Suicide, London, Williams & Wilkins, 97-112, 1986.
6. Hollinger PC: Violent Deaths in the United States. New York, The Guilford Press, 1987.
7. Harrow M, Grinker RR, Silverstein M, Holzman P: Is modern-day schizophrenic outcome still negative? Am J Psychiatry 135: 1156-1162, 1978.
8. Bleuler M: The Schizophrenic Disorders: Long-Term Patients and Family Studies. New Haven and London, Yale University Press, 1978.
9. Buda M, Tsuang MT, Fleming JA: Causes of death in DSM III Schizophrenics and other psychotics (Atypical Group). Arch Gen Psychiatry 45: 283-285, 1988.
10. McGlashan T: The chestnut lodge follow-up study, II: Long-term outcome of schizophrenia and affective disorders. Arch Gen Psychiatry 41: 586-601, 1984.
11. Harding CM, Zubin J, Strauss JS: Chronicity in schizophrenia: Fact, partial fact or artifact? Hosp Community Psychiatry 38: 477-486, 1987.
12. Stephens JH: Long-term prognosis and follow-up in schizophrenia. Schizophr Bull 4: 25-37, 1978.
13. Murphy G: On suicide prevention and prediction. Arch Gen Psychiatry 40: 343-344, 1983.
14. Grinker RR Sr, Holzman PS: Schizophrenic pathology of young adults: A clinical study. Arch Gen Psychiatry 28: 168-175, 1973.
15. Grinker RR Sr, Harrow M: (Eds.) Clinical Research in Schizophrenia: A Multidimensional Approach. Springfield, Charles C. Thomas Press, 1987.
16. Harrow M, Westermeyer JF, Silverstein M, Strauss BS, Cohler BJ: Predictors of outcome in schizophrenia: The process-reactive dimension. Schizophr Bull 12: 195-207, 1986.
17. Summer F, Harrow M, Westermeyer J: Neurotic symptoms in the postacute phase of schizophrenia. J Nerv Ment Dis 171: 216-221, 1983.
18. Westermeyer JF, Harrow M: Prognosis and outcome using broad (DSM II) and narrow (DSM III) concepts of schizophrenia. Schizophr Bull 10: 624-637, 1984.
19. Westermeyer JF, Harrow M: Predicting outcome in schizophrenics and nonschizophrenics of both sexes: The

Zigler-Phillips social competence scale. <u>J Abnormal Psychol</u> 95: 406-409, 1986.

20. Westermeyer JF, Harrow M: Course and outcome in schizophrenia, in <u>Handbook of Schizophrenia</u>, Volume 3: <u>Nosology, Epidemiology and Genetics</u>. Edited by Tsuang, M, Simpson J. Elsevier Science Publishers, 1988.

21. Harrow M, Carone B, Westermeyer JF: The course of psychosis in early stages of schizophrenia. <u>Am J Psychiatry</u> 142: 702-707, 1985.

22. Harrow M, Miller J: Schizophrenic thought disorder and impaired perspective. <u>J Abnormal Psychol</u> 89: 717-727, 1980.

23. Harrow M, Quinlan DM: <u>Disordered Thinking and Schizophrenic Psychopathology</u>. New York, Gardner Press, 1985.

24. Harrow M, Marengo JT: Schizophrenic thought disorder at follow-up: Its persistence and prognostic significance. <u>Schizophr Bull</u> 12: 208-224, 1986.

25. Marengo JT, Harrow M: Thought disorder: A function of schizophrenia, mania, or psychosis? <u>J Nerv Ment Dis</u> 12: 497-511, 1986.

26. Endicott J, spitzer R: A diagnostic interview: The Schedule for Affective Disorders and Schizophrenia. <u>Arch Gen Psychiatry</u> 35: 837-844, 1978.

27. Hollingshead AB, Redlich RC: <u>Social Class and Mental Illness</u>. New York, John Wiley and Sons, 1958.

28. Vaillant GE: A ten-year follow-up of remitting schizophrenics. <u>Schizophr Bull</u> 40: 78-85, 1978.

29. Wechsler D: Weschsler Adult Intelligence Scale Manual. New York, Psychological Corporation, 1955.

30. Breier A, Astrachan B: Characteristics of schizophrenic patients who commit suicide. <u>Am J Psychiatry</u> 141: 206-209, 1984.

31. Tuchman J, Youngman WF: Assessment of suicide risk in attempted suicide. in: Resnick, HLP (Ed) <u>Suicidal Behaviors</u>. Boston, Little, Brown and Co, 1968.

32. Lewine R, Burbach D, Meltzer HY: Effect of diagnostic criteria on the ratio of male to female schizophrenic patients. <u>Am J Psychiatry</u> 141: 84-87, 1984.

33. Drake R, Gates C, Cotton P, Whitaker A: Suicide among schizophrenics: Who is at risk? <u>J Nerv Ment Dis</u> 172: 613-617, 1984.

34. Roy A: Suicide in chronic schizophrenia. <u>Br J Psychiatry</u> 141: 171-177, 1982.

35. Pokorny A: Prediction of suicide in psychiatric patients. <u>Arch Gen Psychiatry</u> 40: 249-257, 1983.

SUICIDE IN SCHIZOPHRENIA: CLINICAL APPROACHES

Robert E. Drake, Stephen J. Bartels
William C. Torrey

Dartmouth Medical School
Hanover, New Hampshire

As part of the current trend in suicide research to focus on specific high-risk groups (1), we have been studying suicide in schizophrenia (2-7). In addition to reviewing the literature (2), we have identified risk factors (3-5), considered the influence of depression and hopelessness (6), and examined the differences between attempters and completed suicides (7). This work, along with that of Roy (8) and others (9-15), enables us to identify high-risk patients rather than to predict specific suicides (16).

Given that approximately 10% of schizophrenic patients complete suicide (17) and that many more make serious attempts (18,19), addressing the suicidal urge in high-risk patients should be a primary goal of treatment. The literature on the treatment of schizophrenia pays little attention to the problem of suicide. We are aware of no studies of high-risk patients. This oversight represents a significant gap in the literature since we know that the majority of young schizophrenic patients consider suicide. For many patients, thoughts about suicide, like thoughts about hospitalization (20), are a primary concern. For the sake of accurate empathy, engagement, and long-term treatment, the patient's thoughts about suicide must be understood and addressed.

In this paper we will summarize briefly the literature on identifying high-risk patients and then discuss potential treatment approaches. We offer these thoughts as clinical hypotheses rather than as guidelines for treatment.

IDENTIFYING HIGH-RISK SCHIZOPHRENIC PATIENTS

Based on reviews by our own group (2) and by Roy (21), risk status can be assessed as follows: Schizophrenic patients who complete suicide tend to be young, male, and unmarried. They tend to have chronic schizophrenia and to kill themselves within the first five to ten years of illness. They typically have a course characterized by exacerbations and remissions. Between psychotic episodes they recover sufficiently to appreciate the devastation that schizophrenia has wrought on their lives and express a painful awareness of their psychopathology. Many have made previous suicide attempts, but suicide attempts are so common among schizophrenic patients as to be relatively unhelpful as predictors.

Suicides in schizophrenia typically occur while the patient is on leave from the hospital or within the weeks or months following discharge from the hospital. Patients who kill themselves tend to live alone and to be socially isolated. They may withdraw from caregivers as well as others just prior to suicide. The most common immediate precipitants are the loss of supportive relationships on leaving the hospital and the loss of familial supports.

A majority of schizophrenics who suicide are depressed at the time of death. Their depressions often meet syndromal criteria for major depressive episodes but are typically characterized by hopelessness about the future. Contrary to traditional clinical wisdom, command hallucinations are rarely associated with suicide. The suicidal drive may, however, be fueled by a state of agitation, akathisia, or a tendency toward impulsive, violent behavior.

As Bleuler (22) recognized long ago, a common motive for suicide appears to be the wish to escape a life that has become unbearable due to the ravages and humiliations of severe and corrosive mental illness. The combination of good premorbid functioning, painful awareness of illness, and inability to maintain previous levels of good functioning led us to designate a "high expectations-awareness" subgroup of high-risk patients (5). Other motivations for suicide certainly exist. For example, some patients kill themselves in the midst of psychotic episodes, some during states of agitation without depression, and others while trying to escape from medication side effects.

Clinical Approaches

There are no controlled studies of the treatment of high-risk schizophrenic patients. Furthermore, the literature does not suggest any clear relationships between treatment and suicide (2,21). Our impressions regarding treatment come from three sources: (a) studies of completed suicides, including psychological autopsies as well as biological investigations; (b) interviews with clinicians who have treated suicidal schizophrenic patients; and (c) our own experience in interviewing schizophrenic patients about their suicidal thoughts, working with some of them to diminish the suicidal urge, and learning from others about how they survived suicidal crises earlier in their lives.

As clinicians who have tried to learn from and help these patients, we believe that treatment can make a difference. Moreover, our clinical experience suggests that focusing on the suicidal urge often facilitates treatment. Many patients are privately absorbed with thoughts of life or death; asking them about suicide and accepting their suicidal preoccupations may be the key to an empathic connection. For example, one young female patient sat silently in a community social club for nearly six months following a lengthy hospitalization and without developing relationships with staff or other patients. Staff attributed her behavior to negative symptoms. When one of us began to ask her about suicide, however, she admitted that she had thought of little else since leaving the hospital. Talking about suicide permitted her to share her despair over the humiliations and many losses occasioned by her illness. Thereafter, we continued to think about suicide and made one attempt over several years but also became progressively engaged in relationships with caregivers and other patients and slowly began to build a new life for herself.

We believe that clinicians should address the suicidal urge when it is present. In what follows we offer suggestions regarding interventions in five domains: hospitalization, somatic treatment, psychotherapy, family therapy and psychosocial rehabilitation.

I. Hospitalization

Because secondary depression and other states of despair and dysphoria associated with suicide are common and occur at various points in the course of schizophrenic illness (23), addressing the suicidal urge should occur throughout treatment. When a suicidal crisis occurs, hospitalization is often helpful. The explicit reason for admission may be suicidal ideation (24), but often the suicidal urge is revealed during hospitalization. Secondary depression appears in approximately one-fourth of schizophrenic patients following the resolution of acute psychotic symptoms (25,26).

When the patient at risk for suicide is identified in the hospital, immediate steps should be taken to assure containment (27), safety, close monitoring, and rapid treatment. When the patient presents an imminent risk or is too withdrawn, agitated, or psychotic to be assessed, a locked unit, locked seclusion room, or one-to-one staff monitoring is indicated. By themselves, these measures may still be inadequate. Suicides occur on inpatient units five to thirty times more frequently than in the general population (28,29). Inpatient suicides among schizophrenics are, however, primarily carried out by patients who have been assessed as clinically improved rather than by those under close observation (30).

For the acutely psychotic patient, recompensation often requires a relatively nonintensive setting (31). Uncovering psychotherapy, confrontational group therapy, and a stimulating milieu are contraindicated. Developing a supportive working relationship that enhances the patient's capacity to request help when overwhelmed by self-destructive impulses should be a primary goal. As described below, somatic interventions should target the specific symptoms of the primary illness and the secondary syndromes.

Evaluating clinical improvement and readiness for discharge is particularly difficult, yet critical. Several recommendations emerge from the empirical data and our own clinical experience. First, suicidal schizophrenic patients may require an extended length of stay that exceeds the resolution of the acute psychosis. Disappearance of active psychotic symptoms should not be the sole criterion for discharge. Symptoms of demoralization, depression, and hopelessness, though frequently unrecognized because they are overshadowed by psychotic symptoms (32), must be evaluated. In reviewing suicides by schizophrenics with clinicians, there are indications that some of these patients were discharged prematurely (3). These patients were in fact requesting an extended hospitalization at the time of discharge. Despite their requests, staff pursued a brief, crisis-oriented approach and expressed concerns that extended hospitalization could result in regression or institutionalization. Countertransference concerns may have been a factor in some of these early discharges. Clinicians in training tend to identify with bright, young, college-educated patients and feel that treatment in public hospitals is inappropriate.

Second, psychosocial assessment and aftercare planning must be early priorities in the hospitalization of the suicidal schizophrenic. Discharge is clearly a time of increased stress. Up to half of all schizophrenic suicides occur within the first few months after discharge, with as many as one-third occurring in the first month (21). The immediate post-hospitalization period requires thorough planning. Adequate material resources, an accepting and supportive social network, and thorough clinical follow-up are critical (33).

Finally, the treatment team and patient must anticipate future suicidal crises and plan for the rapid provision of support, including hospitalization (31). Particularly in the first six months after discharge, a low threshold for rehospitalization is indicated. In addition to frequent clinical assessments, a psychoeducational approach to symptom monitoring that enlists the patient and family members is important (31). Family members are often more able to recognize symptoms of depression or impending relapse than the patient (34). The key to successful inpatient treatment is anticipating and planning for this period of increased vulnerability.

II. Somatic Treatment: Depression and Agitation

We are unaware of any controlled studies of the pharmacological treatment of the suicidal schizophrenic. Nevertheless, a rational approach can be constructed on the basis of studies of the biological factors and behavioral syndromes associated with suicide in schizophrenia. This approach follows the principles for clinical trials outlined by Eichelman (35). The primary disorder should be treated first; secondary syndromes should be addressed thereafter; and trials should consider the risks and benefits of each intervention, should proceed in serial fashion, and should specify measurable target symptoms. In the case of the suicidal schizophrenic, psychosis should be treated appropriately before proceeding to evaluate and treat the two

Table 1. Hospitalization of the Suicidal Schizophrenic

ASSESSMENT:
    Identify the patient at risk

TREATMENT PLANNING:
    Provide containment in safe, low-stimulation
    environment

    Treat primary and secondary symptoms

    Consider the need for longer length of stay

DISCHARGE PLANNING:

    Carefully assess support system and aftercare
    treatment

    Assume low threshold for rehospitalization during
    first 6 months after discharge (when most
    suicides occur)

major predisposing syndromes of depression and agitation (see Table 2).

A. Secondary Depression

    Post-psychotic depression occurs following approximately one-fourth of all schizophrenic episodes (25,26), and approximately sixty percent of all schizophrenics suffer a major depressive episode in the course of their illness (36). Moreover, a variety of studies agree that secondary depression is a major suicide risk factor (2). The majority of schizophrenics are depressed at the time of suicide (6,8), and over 50% have experienced a past episode of depression (8).

    The differential diagnosis of depressive symptoms in schizophrenia is complex (23). The first step is to rule-out underlying substance abuse, medical illness and medication side effects. Neuroleptic-induced akinesia, for example, is quite common, can be indistinguishable from depression, and may require an empirical trial of antiparkinsonian agents for identification (37,38). Covert alcohol abuse is also common and may be associated with depression (39). Among nonpsychotic patients, prodromal symptoms of psychotic relapse, negative symptoms and transient depressive symptoms also must be ruled out (23). For the remaining patients with secondary depressive symptoms, somatic treatments may be effective.

    Symptoms that tend to respond to antidepressants include insomnia, decreased appetite, lack of energy, diminished concentration, subjective sadness, negative self-evaluation and self-reproach (37). Response to antidepressant medications may not be seen for six weeks, and some partial responders will recover more fully after 9 weeks of treatment.

175

We are unaware of studies comparing the efficacy of antidepressants and other somatic treatments for secondary depression. Lithium or electroconvulsive therapy (ECT) may be effective in individual cases. The literature suggests that the treatment of secondary affective syndromes in chronic schizophrenia with ECT produces only transient improvements (40,41), but nonchronic schizophrenic patients with secondary affective symptoms may respond more favorably to ECT (41,42).

B. Secondary Agitation

States of extreme agitation, which may overlap with depressive symptoms, are also associated with suicidal behavior in schizophrenia (5,13). These patients appear tense, may be threatening or assaultive, and report internal states of restlessness, dysphoria, or terror. They are often described as demanding and impulsive, and they sometimes express death imagery in their psychotic ruminations rather than reporting suicidal ideation directly.

Clinicians should first rule out underlying medical conditions which could be causing the agitation. As with the depressive states, substance abuse and medication side effects are common causes of agitation in schizophrenia (4,39). Akathisia in particular has been associated with suicidal behavior in schizophrenia (4,43,44). Patients with episodic suicidal behavior should also be evaluated for temporal lobe dysfunction by electroencephalogram (45).

After addressing possible organic contributions, the agitated, suicidal schizophrenic should be considered for medication trials to relieve the symptoms. Clinical research indicates that suicidal and violent behavior in schizophrenia are closely associated (46-48), and low cerebrospinal fluid levels of 5-hydroxyindoleacetic acid (5-HIAA) have been found in both states (49-51). Although there are no treatment studies of suicidal agitation, studies of the agitated schizophrenic who is violent may help to guide medication choice.

Agitated states associated with violence in schizophrenia have been alleviated using benzodiazepines (52,53), propranolol (54), Lithium (55), tryptophan (56), and anticonvulsants such as carbamazepine (57,58), primidone (59), and clonazepam (60). These medications can be used in systematic empirical trials, using first those agents with the most benign side effects and fewest risks (35). Quantifiable targets such as symptoms of severe anxiety or dysphoria or signs of agitation and violence should be carefully identified and monitored throughout treatment.

III. Psychotherapy

Somatic interventions are often ineffective in the treatment of the depressed, suicidal schizophrenic. In these cases, the suicidal urge seems related to the realistic awareness of having a devastating mental illness. Perhaps,

Table 2.   A Rational Approach to Pharmacotherapy: 2
          Syndromes Associated with Suicide

SECONDARY DEPRESSION

Dx: Hopelessness, Worthlessness, Sadness,
      Decreased: Sleep, Appetite, Energy, Concentration

Rx: 1) Rule-out Akinesia, Substance Abuse, Organic
         Illness, and Other Depression-Like Syndromes

      2) Antidepressants for 6-9 Weeks

      3) ?ECT, Lithium, MAOI'S

SECONDARY AGITATION

Dx: Despairing, Terrified, Labile, Dysphoric, Asxious
      +/-Increased Motor Activity with
      Threatening and Violent Behavior

Rx: 1) Rule-out Akathisia, Substance Abuse, Organic
         Illness

      2) Benzodiazepines, Beta-Blockers, Lithium,
         Tryptophan, Tryptophan + Trazadone,
         Carbamazepine

as Bleuler (22) suggested many years ago, life with
schizophrenia simply becomes an unbearable existence. The
bright, highly motivated, young person is particularly likely
to find the discrepancy between life goals and the reality of
living with severe mental illness intolerable (5).   These
patients are frequently without the vegetative symptoms of
major depression; they suffer instead from a psychological
state of demoralization (6,23,61).

     Whether or not they respond to medications, demoralized
patients need psychological interventions (62). Alleviating
their despair entails helping them to bear the reality of
their life experience (63). Therefore, the psychotherapeutic
approach should be supportive and existential.

     The first and most important step is engaging the
patient. This requires an active and flexible approach to
psychotherapy.   Traditional, office-based psychotherapy,
whether supportive or insight-oriented, leads to high rates
of attrition (64).   Demoralized, suicidal patients are
particularly prone to drop out of treatment (65). Engagement
requires active outreach and a willingness to become involved
in the practical details of the patient's life. The task at
this stage is to support the patient by filling in for and
modeling missing ego deficits rather than interpreting them
(66).

     Once a supportive relationship is in place, the
therapist should be willing to review with the patient his or
her life experience, sometimes in metaphor. At the same
time, the therapist monitors the patient's suicide potential,

not in terms of expressed ideas so much as in relation to the patient's self-esteem (3,67). When self-esteem is further threatened, such as following losses or experiences of failure, the therapist should be willing to increase supports. This may include direct interventions in the patient's life, more frequent visits, or hospitalization.

Over the long-term, psychotherapy must assist the suicidal schizophrenic patient with three tasks: to bear the despair, to mourn the loss of previous expectations, and to establish a new identity. The treatment is largely existential. The therapist should resist the temptation to offer false hope or easy solutions and should instead acknowledge empathically the patient's view that death is one solution to the problem of unbearable psychological pain (68). The therapist demonstrates optimism through an unswerving interest in the patient's welfare, attention to short-term tasks and goals, and addressing the illness and its long-term implications realistically. Support of this type will enable many schizophrenic patients to survive the crisis of despair.

The second long-term goal is to help the patient mourn the loss of former dreams, goals, and self-expectations. For the young, previously high-functioning patient, this is an enormous task. It may be accomplished over time, but the therapist should allow considerable denial and rationalization (69). Complete recovery from demoralization involves relinquishing or seriously modifying previous goals (70), a task which schizophrenic patients should be allowed to pursue at their own pace. The therapist must show great patience and empathy and should allow the patient whatever time is needed for mourning.

The final long-term goal of psychotherapy is to help the patient establish a new identity. This process requires an acceptance of realistic limitations in the context of an appreciation of remaining strengths and current capabilities (71). Obviously, acquiring new skills and the stabilization of illness with time can facilitate this process. As the patient in this phase searches for new activities and relationships in which to find meaning and construct self-esteem, he or she conducts a reappraisal of identity at all levels: social, familial, vocational, and psychological. The therapist, in the terms of existential psychiatry (72), stands with the patient but does not direct his or her progress. The therapist may, however, minimize the further erosion of self-esteem that comes from attempting to attain unrealistic goals. Throughout this lengthy phase of treatment, the therapist seeks to be with the patient empathically and to protect and enhance the patient's self-esteem.

IV. Family Therapy

Families do not cause suicide in schizophrenia, but they can sometimes play a crucial role in prevention. The best

Table 3.   Psychotherapy of the Suicidal Schizophrenic Patient

   APPROACH: SUPPORTIVE-EXISTENTIAL

       SHORT-TERM GOALS

           1)   ENGAGEMENT
           2)   ASSESSMENT
           3)   CRISIS INTERVENTION

       LONG-TERM GOALS

       ASSIST THE PATIENT TO:

           1)   BEAR THE DESPAIR
           2)   MOURN THE LOSS OF EXPECTATIONS
           3)   ESTABLISH A NEW IDENTITY

evidence for this is indirect: Families, like patients, are prone to develop hopelessness in reaction to chronicity, and their withdrawal of support is the most frequent precipitant to suicide (3,10).   Suicidal schizophrenic patients, often burdened by their own unrealistic expectations (5), can also be stressed by familial pressure to accomplish goals that are no longer attainable.   Family treatment can sometimes prevent these reactions in families and thereby preserve and optimize the patient's support system.

Having a schizophrenic relative is enormously stressful. Several authors (73-77) have described the family's reaction to chronic mental illness in terms of stages that include denial, an outpouring of strong feelings as part of the mourning process, and some form of demoralization or despair in the face of chronicaity prior to developing more constructive coping strategies.   As part of what Terkelson (76) calls the "collapse of optimism", families often express despair and sometimes even the wish for their relative's death (73).

Considering these reactions in relation to the patient's internal process of trying to adjust to a devastating illness suggests certain ways that the family can be helpful. Specifically, the family of the suicidal schizophrenic patient can play a crucial role by: a) providing long-term support and preventing isolation; b) creating a noncritical, emotionally benign environment; c) helping the patient to bear the despair of chronicity, mourn the loss of former self, and maintain optimism about the future; and d) modifying their own expectations and helping the patient to develop new goals.   Other responses such as denying the reality of illness, insisting on unrealistically high expectations, losing hope, or abandoning the patient out of frustration may inadvertently reinforce the patient's hopelessness.

Current psychoeducational approaches to working with the families of schizophrenic patients are designed to provide support, education, and specific skills for managing chronic illness (78-80). Although the studies of psychoeducational family interventions have not specifically examined suicidal behavior as an outcome, these therapies clearly aim to protect the patient's living situation and social support system, to decrease the family's sense of helplessness and hopelessness, and to minimize hostility and criticism in the patient's environment. We can infer that all of these should serve to mitigate the patient's suicidal potential.

Table 4. Family Treatment

APPROACH: SUPPORTIVE-PSYCHOEDUCATIONAL

ASSIST THE FAMILY TO:

1) PROVIDE LONG-TERM SUPPORT
2) CREATE A NONCRITICAL, EMOTIONALLY BENIGN ENVIRONMENT
3) FACILITATE THE PROCESS OF BEARING THE DESPAIR AND MOURNING THE LOSS OF FORMER SELF
4) MODIFY EXPECTATIONS AND PROMOTE NEW GOALS

V. Psychosocial Rehabilitation

Enhancing the patient's self-esteem should be the central goal of a psychiatric rehabilitation program. This goal serves to organize the efforts of case management, day hospital treatment, vocational services, and psychiatric care.

To promote self-esteem, the staff of a rehabilitation program should have a patient-oriented attitude (81), should be realistic in their short-term goals, and should maintain hope that patients will improve or recover over the long-term (81,82). They should continually assess what is meaningful and important to the individual patient (82). Understanding the patient's values, experiences and feelings is essential. For example, patients who judge their self-worth on their ability to work or earn money should typically focus on developing vocational skills. For those whose self-worth is strongly tied to their ability to relate with others, communication skills and group work within day treatment can be the central focus of rehabilitation.

All programs tend to have a bureaucratic momentum of their own. Caregivers must be vigilant so that inflexible program-oriented goals, such as full-time employment for all patients, not supplant individual needs (81). Inflexible programmatic goals can actually diminish self-esteem and be harmful for some patients.

Short-term rehabilitation goals must be realistic. Young schizophrenic patients who have significant functional limits can experience increased demoralization if they fail at a task that would have previously been easily achievable (61). Clearly, some patients do not benefit from pressure to learn new skills and become more independent; they function better when allowed a period of increased dependency and a moratorium from expectations (82).

Some schizophrenic patients develop florid psychotic symptoms when too vigorously socially stimulated in rehabilitation settings (83,84). In the vocational arena, other patients find even simple, routine jobs exhausting (85) or impossible. These are dangerous situations for the suicidal young schizophrenic patient who may experience failure as an intolerable assault on self-esteem.

To help patients acquire hope for the future, a rehabilitation staff must themselves develop and maintain an active, hopeful attitude (82) that is consistent with evidence from several long-term follow-up studies (86). Recovered patients indicate that finding a hopeful caregiver allows them to integrate their experience and to begin the recovery process (87).

When psychiatric rehabilitation provides patients with an unhurried, accepting, hopeful environment in which new skills can be learned, it does much more than simply help patients compensate for impairments (88). It also serves as a "holding environment" during the early, chaotic, and most dangerous years of their illness. This "holding" function may be more important than rehabilitation per se; it protects the patient while he or she begins to rebuild a sense of self (88).

Table 5. Psychosocial Rehabilitation

MAY BE HARMFUL IF FOCUSED ON PERFORMANCE

SHORT-TERM GOALS:
    REALISTIC, INDIVIDUALIZED, FLEXIBLE

LONG-TERM GOALS:
    PROVIDE A "HOLDING ENVIRONMENT"
    MAINTAIN A HOPEFUL ATTITUDE
    BUILD SELF-ESTEEM
    FACILITATE REINTEGRATION

VI. SUMMARY

We have attempted to cull from the literature and our own clinical experiences preliminary guidelines for treating the suicidal schizophrenic patient. High-risk patients can be identified on the basis of a number of characteristic risk factors. These patients should be assertively engaged and maintained in treatment. Their suicide potential should be monitored carefully throughout treatment by paying attention to self-esteem as well as to predisposing secondary syndromes. Depression and agitation should be treated with somatic interventions. Clinicians should attempt to protect the patient during periods of increased risk, maintain a hopeful but realistic attitude, provide opportunities for participating in activities and acquiring new skills that will enhance self-esteem, and help the patient and family in the long-term processes of bearing the feelings that attend chronicity, mourning the loss of former expectations, and developing new goals and expectations.

REFERENCES

1. Hendin H: Suicide: a review of new directions in research. Hosp Community Psychiatry 37: 148-154, 1986.
2. Drake RE, Gates C, Whitaker A, Cotton PG: Suicide among schizophrenics: a review. Compr Psychiatry 26: 90-100, 1985.
3. Cotton PG, Drake RE, Gates C: Critical treatment issues in suicide among schizophrenics. Hosp Community Psychiatry 36: 534-536, 1985.
4. Drake RE, Ehrlich J: Suicide attempts associated with akathisia. Am J Psychiatry 142: 499-501, 1985.
5. Drake RE, Gates C, Cotton PG, Whitaker A: Suicide among schizophrenics: who is at risk? J Nerv Ment Dis 172: 613-617, 1984.
6. Drake RE, Cotton PG: Depression, hopelessness, and suicide in chronic schizophrenia. Br J Psychiatry 148: 554-559, 1986.
7. Drake RE, Gates C, Cotton PG: Suicide among schizophrenics: a comparison of attempters and completed suicides. Br J Psychiatry 149: 784-787, 1986.
8. Roy A: Suicide in chronic schizophrenia. Br J Psychiatry 141: 171-177, 1982.
9. Beck AT, Kovacs M, Weissman A: Hopelessness and suicidal behavior: an overview. JAMA 234: 1146-1149, 1975.
10. Breier A, Astrachan BM: Characterization of schizophrenic patients who commit suicide. Am J Psychiatry 141: 206-209, 1984.
11. Cohen S, Leonard CV, Farberow NL, Shneidman ES: Tranquilizers and suicide in schizophrenic patients. Arch Gen Psychiatry 11: 312-317, 1964.
12. Minkoff K, Bergman E, Beckj AT, Beck R: Hopelessness, depression, and attempted suicide. Am J Psychiatry 130: 455-459, 1973.
13. Planansky K, Johnston R: The occurrence and characteristics of suicidal preoccupation and acts in schizophrenia. Acta Psychiatr Scand 47: 473-483, 1971.
14. Shaffer JW, Perlin S, Schmidt CW, Stephens JH: The prediction of suicide in schizophrenics. J Nerv Ment Dis 159: 349-355, 1974.
15. Warnes H: Suicide in schizophrenics. Dis Nerv Sys 29(suppl 5): 35-40, 1968.
16. Pokorny AD: Prediction of suicide in psychiatric patients. Arch Gen Psychiatry 40: 249-257, 1983.
17. Miles CP: Conditions predisposing to suicide. J Nerv Ment Dis 164: 455-459, 1977.
18. Niskanen P, Lonnquist J, Achte KA: Schizophrenia and suicide. Psychiatr Fenn 223-227, 1973.
19. Planansky K, Johnston R: Clinical setting and motivation in suicide attempts. Acta Psychiatr Scand 49: 680-690, 1973.
20. Drake RE, Wallach MA: Mental patients' attitudes toward psychiatric hospitalization: a neglected aspect of hospital tenure. Am J Psychiatry 145: 29-34, 1988.
21. Roy A: Suicide in schizophrenia, in Suicide. Edited by Q. Roy Baltimore, Williams and Wilkins, 1986.

22. Bleuler E: <u>Dementia Praecox or the Group of Schizophrenias</u>. New York, International Universities Press, 1950.
23. Bartels SJ, Drake RE: Depressive symptoms in schizophrenia: a comprehensive differential diagnosis. <u>Compr Psychiatry</u>, in press.
24. Talbott JA, Glick ID: The inpatient care of the chronically mentally ill. <u>Schizophr Bull</u> 12: 129-140, 1986.
25. Mandel MR, Severe JB, Schooler NR et al: Development and prediction of postpsychotic depression in neuroleptic treated schizophrenics. <u>Arch Gen Psychiatry</u> 39: 197-203, 1982.
26. Siris SG, Harmon GK, Endicott J: Postpsychotic depressive symptoms in hospitalized schizophrenic patients. <u>Arch Gen Psychiatry</u> 38: 1122-1123, 1981.
27. Gunderson JG: Defining the therapeutic processes in psychiatric milieus. <u>Psychiatry</u> 41: 327-335, 1978.
28. Farberow NL: Suicide prevention in the hospital. <u>Hosp Community Psychiatry</u> 32: 99-104, 1981.
29. Olin HS: Management precautions for surviving patients after a ward suicide. <u>Hosp Community Psychiatry</u> 31: 348-349, 1980.
30. Sletten IW, Brown ML, Evenson RC, Altman H: Suicide in mental hospital patients. <u>Dis Nerv Sys</u> 33: 328-334, 1972.
31. Drake RE, Sederer LI: Inpatient psychosocial treatment of chronic schizophrenia: Negative effects and current guidelines. <u>Hosp Community Psychiatry</u> 37: 897-901, 1986.
32. Knights A, Hirsch SR: Revealed depression and drug treatment for schizophrenia. <u>Arch Gen Psychiatry</u> 38: 806-811, 1981.
33. Breier A, Strauss JS: The role of social relationships in the recovery from psychotic disorders. <u>Am J Psychiatry</u> 141: 949-955, 1984.
34 Herz M: Prodromal symptoms and prevention of relapse in schizophrenia. <u>J Clin Psychiatry</u> 46: 22-25, 1985.
35. Eichelman B: Toward a rational pharmacotherapy for aggressive and violent behavior. <u>Hosp Community Psychiatry</u> 39: 31
36. Martin RL, Cloninger CR, Guze SB: Frequency and differential diagnosis of depressive syndromes in schizophrenia. <u>J Clin Psychiatry</u> 46: 9-13, 1985.
37. Siris SG, Morgan V, Fagerstrom R et al: Adjunctive imipramine in the treatment of postpsychotic depression. <u>Arch Gen Psychiatry</u> 44: 533-539, 1987.
38. Siris SG, Adam F, Cohen M et al: Targeted treatment of depression-like symptoms in schizophrenia. <u>Psychopharm Bull</u> 23: 85-89, 1987.
39. Drake RE, Osher FC, Wallach MA: Alcohol use and abuse in schizophrenia: a prospective community study. <u>J Nerv Ment Dis</u>, in press.
40. Brandon S, Cowley P, McDonald C et al: Leicester ECT trial: Results in Schizophrenia. <u>Br J Psychiatry</u> 146: 177-183, 1985.
41. Salzman C: The use of ECT in the treatment of schizophrenia. <u>Am J Psychiatry</u> 137: 1032-1041, 1980.
42. Abrams R: <u>Electroconvulsive Therapy</u>. New York, Oxford University Press, 1988.

43. Schulte JR: Homicide and suicide associated with akathisia and haloperidol. Am J For Psychiatry 6: 3-7, 1985.
44. Shear K, Frances A, Weiden P: Suicide associated with akathisia and depot fluphenazine treatment. J Clin Psychopharm 3: 235-236, 1983.
45. Tucker GJ, Price RP, Johnson VB, McCallister T: Phenomenology of temporal lobe dysfunctions: a link to atypical psychosis - a series of cases. J Nerv Ment Dis 174: 348-356, 1986.
46. Convit A, Jaeger J, Lin SP et al: Predicting assaultiveness in psychiatric inpatients: a pilot study. Hosp Community Psychiatry 39: 429-434, 1988.
47. Krakowski M, Volavka J, Brizer D: Psychopathological violence: a review of the literature. Compr Psychiatry 27: 131-148, 1986.
48. Tardiff K, Sweillam A: Assault, suicide and mental illness. Arch Gen Psychiatry 37: 164-169, 1980.
49. Mann JJ: Psychobiologic predictors of suicide. J Clin Psychiatry 48(12, Suppl): 39-43, 1987.
50. Ninan PT, van Kammen DP, Scheinen M et al: CSF 5-hydroxyindoleacetic acid levels in suicidal schizophrenic patients. Am J Psychiatry 141: 566-569, 1984.
51. van Praag HM: CSF 5-HIAA and suicide in non-depressed schizophrenics. Lancet 2: 977-978, 1983.
52. Kalina RK: Diazepam: its role in a prison setting. Dis Nerv Sys 25: 101-107, 1964.
53. Salzman C, Green AL, Rodriguez-Villa F, Jaskin G: Benzodiazepines combined with neuroleptics in the management of severe disruptive behavior. Psychosomatics 27: 17-21, 1986.
54. Sorgi PJ, Ratey JJ, Polakoff S: Beta-adrenergic blockers for control of aggressive behaviors in patients with chronic schizophrenia. Am J Psychiatry 143: 775-776, 1986.
55. Tupin JP, Smith DB, Classon TL et al: Long-term use of lithium in aggressive prisoners. Compr Psychiatry 14: 311-317, 1973.
56. Morand C, Young CN, Ervin FR: Clinical response of aggressive schizophrenics to oral tryptophan. Biol Psychiatry 18: 575-578, 1983.
57. Hakola HP, Laulumaa VA: Carbamazepine in the treatment of violent schizophrenics. Lancet 1: 1358, 1982.
58. Klein E, Bental E, Lerer B et al: Carbamazepine and haloperidol vs. placebo and haloperidol in excited psychoses. Arch Gen Psychiatry 41: 165-170, 1984.
59. Monroe RJ: Anticonvulsants in the treatment of schizophrenia. J Nerv Ment Dis 160: 119-126, 1975.
60. Keats MM, Mukherjee S: Antiaggressive effect of adjunctive clonazepam in schizophrenia associated with seizure disorder. J Clin Psychiatry 49: 117-118, 1988.
61. Klein DF, Gittleman R, Quitkin F, Rifkin A: Diagnosis and Drug Treatment of Psychiatric Disorders: Adults and Children. Baltimore, Williams and Wilkins, 1980.
62. Frank JD: Persuasion and Healing. New York, Schocken Books, 1974.
63. Semrad EV: Teaching Psychotherapy of Psychotic Patients. New York, Grune and Stratton, 1969.
64. Gunderson JG, Frank AF, Katz HM et al: Effects of psychotherapy in schizophrenia: II. Comparative outcome

of two forms of treatment. <u>Schizophr Bull</u> 10: 564-598, 1984.

65. Virkkunen M: Attitude toward psychiatric treatment before suicide in schizophrenia. <u>Br J Psychiatry</u> 128: 47-49, 1976.

66. Harris M, Bergman HC: Case management with the chronically mentally ill: a clinical perspective. <u>Am J Orthopsychiatry</u> 57: 296-302, 1987.

67. Buie DH, Maltsberger JT: <u>The Practical Formulation of Suicide Risk</u>. Cambridge, Ma, Firefly Press, 1983.

68. Havens L: <u>Making Contact: Uses of Language in Psychotherapy</u>. Cambridge, Ma, Harvard University Press, 1986.

69. Lamb HR: Young adult chronic patients: the new drifters. <u>Hosp Community Psychiatry</u> 33: 465-486, 1982.

70. Bibring E: The mechanism of depression, in <u>Affective Disorders: Psychoanalytic Contribution to their Study</u>. Ed. Greenacre P, New York, International Universities Press, 1953.

71. Weisman A: <u>The Coping Capacity: On the Nature of Being Mortal</u>. New York, Human Sciences Press, 1984.

72. Heavens L: <u>Approaches to the Mind</u>. Boston, Little, Brown, Inc. 1973.

73. Group for the Advancement of Psychiatry: Report No. 119. A Family Affair. Helping Families Cope with Mental Illness: A guide for the Professions. New York, Brunner/Mazel, 1986.

74. Raymond ME, Slaby AE, Lieb J: Familial response to mental illness. <u>Social Casework</u> 56: 492-496, 1975.

75. Smyer MA, Birkel RC: Interventions with families of the chronically mentally ill elderly, in <u>The Chronically Mentally Ill Elderly: Directions for Research</u>. Ed. Light E, Lebowitz B. Washington, DC, Government Printing Office, in press.

76. Terkelson KG: The evolution of family responses to mental illness through time, in <u>Families of the Mentally ILL: Coping and Adaptation</u>. Ed. Hatfield AB, Lefley HP, New York, Guilford, 1987.

77. Tessler RC, Killian LM, Gubman GC: Stages in family response to mental illness: an ideal type. <u>Psychosoc Rehab J</u> 10: 3-16, 1987.

78. Anderson CM, Reiss DJ, Hogarty GE: <u>Schizophrenia and the Family: A Practitioner's Guide to Psychoeducation and Management</u>. New York, Guilford Press, 1986.

79. Bernheim KF, Lehman AF: <u>Working with Families of the Mentally Ill</u>. New York, W.W. Norton, 1985.

80. Falloon IRH, Boyd JL, McGill CW: <u>Family Care of Schizophrenia</u>. New York, Guilford Press, 1984.

81. Wing JK: Rehabilitation and management of schizophrenia, in <u>Handbook of Psychiatry 3; Psychosis of Uncertain Aetiology</u>. Edited by Wing JK, Wing L. New York, Cambridge University Press, 1982.

82. Anthony WA, Cohen MR, Cohen BF: Philosophy, treatment process, and principles of the psychiatric rehabilitation approach, in <u>Deinstitutionalization</u>. Edited by Bachrach L, San Francisco, Jossey-Bass, 1983.

83. Stone AA, Eldred SH: Delusional formation during the activation of chronic schizophrenic patients. <u>Arch Gen Psychiatry</u> 1: 177-179, 1959.

84. Wing JK, Bennett DH, Denham J: The industrial rehabilitation of long-stay schizophrenic patients, in

<u>Medical Research Council Research Memorandum No. 42</u>
London, HMSO, 1964.

85. Hewett S, Ryan P, Wing JK: Living without the mental
    hospitals. <u>J Soc Policy</u> 4: 391-404, 1975.

86. Harding CM, Zubin J, Strauss JS: Chronicity in
    schizophrenia: fact, partial fact, or artifact? <u>Hosp
    Community Psychiatry</u> 38: 477-486, 1987.

87. Lovejoy M: Recovery from schizophrenia: a personal
    odyssey, in <u>The Chronic Mental Patient/II</u>. Edited by
    Menninger WW, Hannah G, Washington, American Psychiatric
    Press, 1987.

88. Struass JS: Discussion: what does rehabilitation
    accomplish? <u>Schizophr Bull</u> 12: 720-723, 1986.

Section 4

TREATMENT APPROACHES

INTRODUCTION

G.M. McDougall and D. Vosburgh

Treatment approaches integrate theoretical aspects with our ability to classify the phenomenology and pathophysiology of any given condition. In this section, relating to treatment approaches of depression in schizophrenia, the presentations outline the complexity and uncertainty of our understanding of each of these phenomena--schizophrenia and depression. For a variety of reasons many of the basic definitions have not been agreed upon with either of these diagnostic categories. Nor is there a full understanding of the profound interference and side-effects of medications used in treatment. This being the case, it is obvious that treatment will remain uncertain until more is known. However, the papers within this section explore those areas which clinicians inquire about. Is it the schizophrenic illness? is there a true depression? Is what is seen a side effect of medication? will an antidepressant be effective?

Dr. D.A.W. Johnson clearly indicates that there is no single etiology and therefore no single treatment or management. His paper provides an overview of the complexity of the diagnosis and treatment of depression in schizophrenia. He reviews the background in terms of personality and genetic predisposition, and notes that the condition exists in a person in an environment, not in a vacuum. He elaborates on four potential ways which depression may be part of the schizophrenic picture: an ordinary part of the illness as described by Bleuler; part of a post-psychotic depression; related to neuroleptic drug interference; or something ubiquitously called the schizoaffective illness.

Following Dr. Johnson's general overview, the contributors present more specific aspects of their interest in this area. Dr. B.D. Jones presents a hypothesis that there are three mechanisms by which depression or depressive symptoms can develop in response to long term neuroleptic exposure or toxicity. He terms the first 'disfigurement',a depression as a response to tardive dyskinsia. The second mechanism is a depression which is a 'physiological dysfunction' secondary to the presence of a tardive syndrome, and he speculates that there is a third depression possibly resulting from actual 'damage' or altered brain functioning. He concludes with the challenge to seek new ways to study schizophrenia so that neuroleptic artifact can be avoided. This challenge is to understand the nature of schizophrenia and its varied manifestations including depression, and to eliminate the confusion as to which is illness or treatment related.

Dr. S.G. Siris reports on the use of antidepressants in schizophrenics with depression. He outlines the limited number of prospective double blind controlled trials which supplement an antidepressant to the neuroleptic treatment regime of schizophrenic patients. In the studies reviewed, the effectiveness of the use of antidepressants has been mixed. He indicates that the diagnosis of depression is often uncertain, but the symptoms are those resembling the clinical picture of depression. However, Dr. Siris suggests guidelines for treatment based on our current understanding. With the development of depression-like symptomatology, the first step should not to immediately introduce an antidepressant. As Dr. Johnson concurred, observation and review of the patient should start with reassessment and reinforcement of the patient's biopsychosocial support system. Dr. Siris suggests that if the new sympomatology is part of an incipient psychotic decompensation, perhaps additional neuroleptic medication is necessary. Conversely if the picture represents neuroleptic side effects, the patient's state of well being may be improved by a reduction in neuroleptic dosage . He provides guidelines for trials of antiparkinson and antidepressant medications during certain times of the depressive illness.

Dr. T. Van Putten, S.R. Marder, and N. Chabert are concerned with the effects of medication dysphoria, particularly at higher dosages. They believe that akathisia is significantly related to depression in those who are treated with neuroleptic medications. They propose that dysphoric responses to medication erode the patient's quality of life, and that schizophrenics often correctly attribute their dysphoric state to their medications. They believe that many of these dysphoric feeling states are extrapryamidally based, are dose dependent, and that a reduction in dosage reduces a neuroleptic induced dysphoria.

Drs R. Williams and J.T. Dalby report on drug and subject influences on psychometric measures of depression and negative symptoms in schizophrenic patients. Their study examines the phenomenology of the schizophrenic syndrome and attends to the differential usefulness of measurement instruments of depression. They reject the hypotheses that depression may be caused by neuroleptics or that depression is a manifestation of 'depression-like' negative symptoms, and reiterate Kraeplin and Bleuler's concern with 'genuine' depression in the conditions called the schizophrenias.

This section on the treatment of the schizophrenic patient provides an overview, theories on causation, and considerations of current practice. Further advances with newer neuroleptic medications may soon clarify how much of the depression-like symptoms are actually drug interference. Other treatment approaches will develop as we increase our knowledge and understanding of the etiology (genetic, neuropathological, neurochemical and environmental) of the schizophrenic disorders. More specific treatments depend on these advances.

THE COMPLEX PROBLEM OF TREATMENT

D.A.W. Johnson

University Hospital of South Manchester
Manchester, England

Although the presence of depressive mood disorders in schizophrenia has been recognised since the days of Kraepelin and Bleuler the prevalence of mood disorders and the resultant morbidity has been assessed only recently. A review of the literature suggests a frequency of 15-50 percent over 9-24 months (Table 1).

Individual studies highlight a number of important results. Both the McGlashan and Carpenter (17,18) and Knights and Hirsch (16) papers suggest a particularly high risk in the postpsychotic period (up to 50 percent). However, perhaps it is the discovery of the morbidity caused by affective symptoms that has been the more surprising result. Cheadle et al (3) in a comprehensive study of outpatients maintained in the community found that the most frequent symptoms complained of by these patients were of anxiety and depression (57 percent). Falloon et al (6) reported that depression gives an equal risk of hospital admission as acute psychotic symptoms. Johnson (13), using particularly strict criteria with respect to both duration and severity, found affective symptoms twice as frequent as acute positive symptoms, but the consequences to the patient with regard to work and social function, and the risk of admission were considerably less. In schizophrenia the risk of suicide is high with the best estimates suggesting that 10-15 percent of all schizophrenics kill themselves. The exact relationship between suicide and clinical depression remains uncertain but reviews of the literature suggest an association, with a particular risk during the post-psychotic period.

It is most unlikely that a single aetiology will explain all depression, and even individual patients may have different causes on different occasions. This presents a

Table 1   Frequency of Depression in Schizophrenia

Year	Study	%
1967	Helmchen and Hippius	50
1973	Hirsch et al	15 (9 months)
1976	McGlashan and Carpenter (review)	25 (PPD)
1976	McGlashan and Carpenter	50 (PPD)
1977	Cheadle et al (Salford Register)	57
1978	Falloon et al	40
	Admissions: Depression = 16%	
	Schizophrenia = 13%	
1981	Knights and Hirsch	54
1981	Johnson	
	1st Illness: No drugs	
	Prodrome	29
	Acute phase	19
	Chronic illness at relapse	
	No drugs	30
	On drugs	38
	Remission on drugs	
	Antidepressants prescribed	15
	Psychiatrist assessment	25
	Nurse + self-assessment	26

NB PPD = Postpsychotic depression

complex problem when we have to consider the treatment and management of patients with depressive mood disorders. There can be no simple or single approach to the treatment, and on each occasion the full range of possibilities must be considered.

PERSONALITY

Illness does not exist in a vacuum but in a person. The pre-morbid personality with its strengths and vulnerabilities cannot be ignored. The relevance of personality development, life-events, pre-illness behaviour patterns and previous depressions needs to be explored. Roy et al (21) concluded that depressed schizophrenic patients had a more vulnerable personality and experience more stress than non-depressed patients. Depressed schizophrenic patients were more likely to be living alone, have suffered early parental loss, been treated for a previous depression, had previous attempts at self-harm, had more undesirable life-events in the previous six months and had more hospital admissions.

GENETIC PREDISPOSITION

It has been suggested that patients who have first degree relatives with affective disorders are more at risk to experience depression in the course of their schizophrenic illness. Even if this is correct, there is currently no information on the natural history or treatment response of these depressions.

Alternatively Galdi et al (7) suggest that 'pharmacogenic depression' occurs in genetically predisposed patients. In their opinion depression represents an extrapyramidal component of a dopaminergic-related disorder.

POSTPSYCHOTIC DEPRESSION

There is a general consensus that the months following recovery from an acute episode of schizophrenia is a period of high risk for all types of morbidity. McGlashan and Carpenter (17) reviewed the literature and concluded that overall 25 percent of patients experienced depression in the first six months after recovery from an acute psychosis.

This is supported by the NIMH study (Mandel et al 19) and the observation by Roy (20) that 50 percent of suicides occurred within three months of discharge.

The theories of aetiology fall into two principal categories. The theories of psychological development or reaction to the experience of psychosis and the suggestion that depression occurring at this time is only an apparent phenomenon due to a differential resolution of existing symptoms ('revealed depression').

The arguments for a psychological development or reaction to the experience of psychosis remain based upon personal observations and anecdotal accounts. Despite the age of these theories they remain unsupported and untested by prospective research studies. The clinical syndromes described would appear to have no consistent features. In contrast 'revealed depression' has been subjected to prospective analysis (16,17,18) and debated extensively (8,11). It is likely that 'revealed depression' during the postpsychotic period explains at least some of the depression reported in this period. However, in considering treatment it is crucial to note that both theories of postpsychotic depression suggest that it is an index of progress and resolution.

Recent research by Johnson (15) reported that the development of postpsychotic depression had no correlation with the risk of a further acute schizophrenic relapse within the next three years, in contrast to a positive correlation with depression at other times. On this basis it was suggested that aetiologically postpsychotic depression should be considered a separate entity.

RELATIONSHIP WITH NEUROLEPTIC DRUGS

Hirsch (10) reports that over 30 papers have been published supporting the argument that depression is due to neuroleptic medication. It is possible for drugs to be associated with depressive symptoms in one of two ways.

(a)   Neuroleptic drugs produce a true pharmacogenic depression. There are many reasons for believing that even if neuroleptic drugs can cause depressions on occasions, this must be a minority cause. Historically depression was recorded both as a prodrome to a first schizophrenic illness, and as an intrinsic part of the illness before the use of neuroleptic drugs (4). In more recent times it is reported as a prodrome to schizophrenic illness, before the initial prescription of neuroleptic drugs, in over 20 percent of patients (13). In double-blind trials not only has depression recurred in the placebo group, but the prevalence has been significantly higher (12). The concept of 'revealed depression' has previously been discussed. It is important to emphasize that depression in these post-psychotic patients resolves with continuing neuroleptic medication. Twice as many patients have full symptom resolution as those continuing with symptoms (19). Johnson (13) studied the prevalence in several different patient groups, the lowest frequency of depression was recorded in those patients receiving regular maintenance medication (Table 1). It must also be noted that no relationship has been found between suicide and neuroleptic medication.

A number of observations suggest that a minority of patients may have a true pharmacogenic depression. The issue of dose related depressions is dealt with fully elsewhere in this volume. A number of authors have reported a higher incidence of depression amongst patients with side-effects including overt extra-pyramidal symptoms. Johnson (13) analysed his dose prescriptions and found a significant difference in the risk of depression only when comparing the very highest doses with the very lowest doses. The recent micro-dose studies have not studied depression systematically, but Johnson (13) found no relationship with dose, although it was the most frequent single symptom present before an acute relapse. The possibility of different neuroleptics having a different effect on mood has been explored without any firm conclusion. Some studies suggest flupenthixol may possess a mood elevating property, and trials have demonstrated an antidepressant action when prescribed in doses lower than the usual anti-psychotic range.

(b)   A number of authors have identified 'akinetic' depression, which is a drug induced neurological or psychological syndrome of extrapyramidal origin, so similar to true depression that either the patient or therapist can confuse the diagnosis (22). Negative features of schizophrenia may further complicate the correct identification of this syndrome. However, for the purposes of considering treatment there are two important facts. The first is that these patients can be identified by a careful consideration of muscle symptoms, and the second is that this group of patients show a rapid response to anticholinergic drugs.

SCHIZO-AFFECTIVE ILLNESS

The possibility of a separate diagnostic category has been discussed by Professor Brockington. It is fair to conclude that at the present time there is no general agreement on the definition of schizo-affective disorder, nor is there prospective evidence that any one of these definitions has a different natural history or outcome from the others. Two drug trials testing the treatment response of 'schizo-affective' patients found that lithium is as helpful as neuroleptics in schizo-manic patients, but there was a better response to chlorpromazine than amitriptyline in schizo-depressive patients. On the basis of these results it was suggested that many schizo-manic patients suffer from mania, but schizo-depressive patients are more likely to have schizophrenia (2).

DEPRESSION JUST ANOTHER SYMPTOM OF SCHIZOPHRENIA

Historically depression was described a frequent symptom of schizophrenia, but more especially as part of the prodrome of a first illness (4). More recent studies have confirmed the presence of depression both in the prodromal phase and acute illness. Depression is equally frequent in the drug treated and drug free patient. Over a 2-3 year period no less than 30 percent of patients will experience this syndrome, and if the period of observation includes an acute relapse the frequency will be 50 percent or higher. Some recent work has suggested that the identification of depression as a new symptom in a stabilized outpatient may be an important index of deterioration, and as a consequence offers the possibility of a new drug treatment strategy to reduce the total dose prescribed in maintenance therapy. The very frequency and broad distribution of the symptom must itself be a cogent argument for depression to be part of the wider syndrome. The recent suggestion of a positive relationship between impending relapse, or risk of future relapses over a 3 year period, must be an even stronger argument for this aetiology in many patients. The resolution of depression with the continuing improvement of other symptoms, particularly in the postpsychotic period, and the survey reports showing that the lowest prevalence of depression is in stabilized patients living in the community on regular maintenance medication must further argue for depression as a symptom of schizophrenia. The drug trial results demonstrating response to neuroleptics adds further weight to this conclusion.

TREATMENT

Since there is no single agreed aetiology, and almost certainly there is no single cause for all depression - even in the same patient - there can be no single treatment.

However, guidelines can be obtained from the literature for a practical approach in our normal clinical practice.

The first option is to offer no additional or different treatment, but to monitor the patient carefully. For depression occurring within the first six months after the apparent resolution of the acute psychotic or positive symptoms, approximately one-half of all depressive symptoms will clear within a two month period. There will be a substantial clinical relapse in less than 10 percent of patients within this interval. Perhaps this outcome should not be unexpected since most theories agree that the emergence of depression at this time is an index of improvement, whatever the theory of aetiology. In depression which develops in patients who have been stable in the community on regular maintenance medication depression will resolve in one-third of patients, and a further 20 percent will have experienced a significant clinical deterioration of a schizophrenic-type within a two month period. These results argue for caution and a degree of patience before instituting new treatment strategies since either the symptoms will resolve, or the need for a more positive intervention will become unequivocally apparent within a two month period. However, a disinclination to rush into alternate active treatments must not represent complacency or be an excuse for not reviewing the full clinical status of the patient since the risk of suicide or harm to the patient is high at this time. The final decision on whether the patient can be left without a more positive intervention will ultimately depend on clinical judgment based on a number of considerations.

At all times the emergence of depression during periods of extended remission must raise the question as to whether the patient is undergoing a schizophrenic deterioration. In a number of studies affective symptoms have been suggested as the single most common symptom in the prodrome to an acute relapse. The risk after the appearance of depression in the next two months is reported as 20 percent, but this rises to 30-35 percent over the next 12 months and to 60 percent over the next 2-3 years, with a more relapsing course over the 3 year period monitored prospectively (19,15). The treatment of a relapsing schizophrenic is, of course, a matter of clinical judgment and may include any of the usual therapies of social, psychological or physical type.

Next the possibility of 'akinetic' depression must be carefully considered. The frequency of this syndrome will vary with the nature of the patient population studied since it will depend on the dose and type of neuroleptic used. In drug resistant illnesses, acutely ill patients and inpatients in general the frequency is likely to be higher. On the basis of a double-blind trial of orphenadrine in depressed patients Johnson (13) estimated that in stable outpatients on depot neuroleptics alone that 10-15 percent of patients may be suffering from this syndrome. On careful questioning most patients will describe muscle sensation changes, most frequently a weakness or stiffness. The syndrome responds rapidly to an intramuscular challenge test with anticholinergic drugs, or oral medication over one week. However, if at all possible the longer term treatment of this condition should be by a dose reduction of neuroleptics, or the use of neuroleptic drugs with a reduced risk of extrapyramidal symptoms. It was anticipated that the micro dose trials would report a lower incidence of extrapyramidal syndromes in general, but the trials published so far have not confirmed this expectation after an initial reduction in the early months.

A full clinical and social evaluation is required as a standard management procedure in the presence of new symptoms. It must be remembered that the presence of any stress or life-event may result in either a direct affective response, or the indirect response of activating schizophrenia. In either event an appropriate physical treatment may be required in addition to attention to resolution of environmental factors. A careful search must always be made for alcohol or substance misuse, this should include careful enquiry about the use of anticholinergic drugs. True pharmacogenic depression must at least be considered. In the absence of other possibilities in an otherwise stable patient a reduction of neuroleptic dose may be tried, or a change of neuroleptic considered. The recent literature increasingly emphasizes that higher neuroleptic doses are unlikely to provide an increased therapeutic response in general, although it may be true for individuals. Many patients routinely receive too high a maintenance dose. Although still a controversial issue, it should be noted that flupenthixol has been suggested as less likely to cause depression for the reasons discussed.

Surveys of drug prescriptions to schizophrenic patients reveal that in any one year 20 percent are likely to receive tricyclic antidepressants in addition to other medication, with a point prevalence of 12 percent. The literature on the response of 'depressed' schizophrenic patients to antidepressants will be reviewed elsewhere in this volume by Dr. S. Siris. However it must be clearly understood that despite the frequency of the prescription of these drugs the practice remains controversial. There are few prospective double-blind trials evaluating the prescription of tricyclic antidepressants to patients on continuing maintenance medication who become depressed. One such trial in the postpsychotic period gives a positive result and another in stable outpatients gives a negative result. Only one trial explores the combination as a form of longer term continuing therapy, and this suggested a benefit to a minority of patients at the expense of increased thought disorder and a very high drop-out rate. The prescription of antidepressants should only be done on an individual basis and in the full understanding that in our present state of knowledge it should be in the nature of a therapeutic trial of medication that requires both careful consideration and monitoring. The risks of possible increased side-effects and a possible deterioration in the underlying schizophrenia must be fully appreciated.

The place of lithium in the treatment of schizophrenia remains unclear with conflicting trial results. Some studies have suggested a benefit to patients with recurrent mood disorders that extends beyond the control of the mood disorder. Others report only that the affective symptoms benefit. More recent work has suggested that lithium may have a therapeutic effect in resistant schizophrenia even in the absence of major mood disorders. Delva and Letermendia (5) have recently reviewed the literature and conclude that its use is probably justified in problem cases even though the issues are not yet satisfactorily resolved.

The use of ECT may be rarely required in patients with severe resistant depression, particularly if there is a risk to life. In selected cases the immediate response to ECT can be rewarding, but there is no evidence it has a lasting effect on the course of schizophrenia.

It must be remembered that the lowest prevalence of depression in chronic schizophrenia is amongst well controlled patients on modest doses of maintenance neuroleptic medication. There is no suggestion of a different risk with oral or depot injection methods of drug administration, nor would it make any sense that this should be so. It is likely that the best treatment is prophylaxis with adequate long term neuroleptic medication in minimal doses that are personalized and kept under constant review in association with the full range of social and psychological strategies.

REFERENCES

1. Brockington I F, Kendell R E, Wainwright S, Hillier VF, Walker J: The distinction between affective psychoses and schizophrenia. Br J Psychiatry 125: 243-248, 1979.
2. Brockington I F, Leff JP: Schizo-affective psychosis: definitions and incidence. Psychol Med 9: 91-99, 1979.
3. Cheadle AJ, Freeman HL, Korer J: Chronic schizophrenic patients in the community. Br J Psychiatry 132: 211-227.
4. Conrad K: Die beginnende schizophrenia: versuch einer gestaltanalyse des wahns. Thieme: Stuggart. p. 315, 1958.
5. Delva NJ, Letemendia FJJ: Lithium treatment in schizophrenia and schizo-affective disorders, in Contemporary Issues in Schizophrenia, Edited by Kerr A, Snaith P. Gaskell, London 381-396, 1986.
6. Falloon O, Watt DC, Shepherd M: A comparative controlled trial of pimozide and fluphenazine decanoate in the continuation therapy of schizophrenia. Psychol Med 8: 59-70, 1978.
7. Galdi J, Rieder RO, Silber D, Bonato RR: Genetic factors in the response to neuroleptics in schizophrenia: a psychopharmacogenetic study. Psychol Med 11: 713-728, 1981.
8. Galdi J: Depression 'revealed' in schizophrenia: A discussion, in Contemporary Issues in Schizophrenia, ed. Kerr A, Snaith P, Gaskell, London 462-466, 1986.
9. Helmchen H, Hippius H: Depressive syndrome im verlauf neuroleptischer therapie. Nervenarzt 38: 455-458, 1967.
10. Hirsch SR: Depression in schizophrenia, in Schizophrenia, Edited by Hirsch SR, Update Publications, London, 1984.
11. Hirsch SR: Depression 'revealed' in schizophrenia, in Contemporary Issues in Schizophrenia, Edited by Kerr A, Snaith P, Gaskell, London, 459-461 and 467-469, 1986.

12. Hirsch SR, Gaind A, Rohde PD, Stevens BC, Wing JK: Outpatient maintenance of chronic schizophrenic patients with long-acting fluphenazine: a double-blind placebo trial. Brit Med J, i, 633-637, 1973.
13. Johnson DAW: Studies of depressive symptoms in schizophrenia.
    i)   Prevalence of depression and its possible causes.
    ii) A two year longitudinal study of symptoms.
    iii) A double-blind trial of orphenadrine against placebo.
    iv) A double-blind trial of nortriptyline for depression in chronic schizophrenia.
    Br J Psychiatry 139: 89-101, 1981.
14. Johnson DAW, Ludlow JM, Street K, Taylor RDW: Double-blind comparison of half-dose and standard-dose flupenthixol decanoate in the maintenance treatment of stabilized out-patients with schizophrenia. Br J Psychiatry 151: 634-638, 1987.
15. Johnson DAW: The significance of depression in the prediction of relapse in chronic schizophrenia. Br J Psychiatry 152: 320-323, 1988.
16. Knights A, Hirsch SR: Revealed depression and drug treatment for schizophrenia. Arch Gen Psychiatry 38: 806-811, 1981.
17. McGlashan TH, Carpenter WT: Postpsychotic depression in schizophrenia. Arch Gen Psychiatry, 33: 231-239, 1976a.
18. McGlashan TH, Carpenter WT: An investigation of the postpsychotic depressive syndrome. Am J Psychiatry, 133: 14-19, 1976b.
19. Mandel MR, Severe JB, Schooler NR, Gelenberg AJ, Mieske M: Development and prediction of postpsychotic depression in neuroleptic treated schizophrenics. Arch Gen Psychiatry, 39: 197-203.
20. Roy A: Suicide in chronic schizophrenia. Br J Psychiatry, 141: 171-177, 1982.
21. Roy A, Thompson R, Kennedy S: Depression in schizophrenia. Br J Psychiatry 142: 465-478, 1983.
22. Van Putten T, May PRA: Akinetic depression in schizophrenia. Arch Gen Psychiatry 35: 1101-1107, 1978.

LONG-TERM NEUROLEPTIC TOXICITY AND MOOD:

BLURRING OF DIAGNOSTIC BOUNDARIES

Barry D. Jones

Department of Psychiatry, University of Ottawa
and Royal Ottawa Hospital Ottawa, Canada

Exposure to neuroleptics results in long-term toxic syndromes, the most well described being tardive dyskinesia. Although schizophrenic patients are most prone to develop tardive dyskinesia because they are exposed to neuroleptics more frequently and for longer duration, evidence suggests that patients suffering from mood disorders may have a significant propensity towards developing tardive dyskinesia (1-11). The converse of this hypothesis, whether long-term neuroleptic exposure can in turn lead to symptoms resembling a mood disorder has been less extensively studied (12). In other words, can a patient, independent of diagnosis, develop long-term neuroleptic toxicity that will result in depression, lability of mood and other symptoms usually associated with a depressive syndrome. This paper will examine this hypothesis from the perspective of three types of mechanisms by which depression or depressive symptoms could develop in response to long-term neuroleptic toxicity. These three mechanisms are: 1) disfigurement with depression as a secondary reaction psychologically to the development of a tardive syndrome such as disfiguring tardive dyskinesia; 2) dysfunction with depression as a secondary effect physiologically to the development of a tardive syndrome such as respiratory dyskinesia or self-induced water intoxication syndrome (SIWIS); and 3) damage with depression, the result of a primary direct toxic effect putatively induced through damage to the frontal lobes.

The possibility that long-term neuroleptic exposure can result in mood disorder symptoms has a number of research implications but more importantly may distort the original presentation of the illness resulting in a blurring of diagnostic boundaries between mood disorder and schizophrenia.

<u>Disfigurement with Depression as Secondary Reaction</u>
<u>Psychologically to Development of a Tardive Movement Disorder</u>

Suicide and depression in tardive dyskinetic patients has been studied in a number of manners. Suicide has been ascribed directly to the embarrassment caused by tardive dyskinetic movements (13,14). In some cases this has been related to direct physical disfigurement resulting from tardive dyskinesia. The difficulty in this regard is to differentiate the risk of suicide related to tardive dyskinesia and disfigurement from the risk of suicide due to the illness for which neuroleptic treatment was originally prescribed. Of interest is a reported case by Chouinard and colleagues (15) of a male who developed severe tardive dyskinesia after treatment with neuroleptic drug for gastrointestinal disturbance. Previously there had been no psychiatric history. Subsequently this patient developed symptoms resembling depression including suicidal ideation due to severe tardive dyskinetic movements in the facial region.

Disfigurement with secondary depression need not only result from tardive dyskinesia. Another tardive syndrome that results in disfigurement and therefore can be expected to be associated with reactive depression is tardive dystonia (16). Tardive or chronic dystonias are not productive of movements but produce bizarre abnormal postures. Although the postural disturbance can be as minor as abnormal extension of the fingers while walking it can be severe involving the truncal musculature. We have had direct experience with patients suffering from tardive dystonia of the truncal musculature which has lead them to be shunned and ostracized by fellow patients resulting in depression and social withdrawal.

Tardive akathisia is another tardive syndrome which may result in a form of disfigurement (17,18). Acute akasthia represents the subjective experience of motor restlessness combined with movements of the lower extremities and occasionally other muscle groups. Both violence and suicide have been associated with the acute forms of akathisia presumably due to the extreme distress subjectively. Tardive or chronic akathisia refers to a chronic form of the disorder where the movements of the lower extremities resemble those seen in acute akathisia with subjective distress but do not seem related to recent increases in dose of medication. Constant movement of muscle groups such as lower extremities in this fashion may cause the patient to look restless, nervous and agitated. The movements may be commented upon by friends and family and ultimately lead to embarrassment and corresponding depression or withdrawal. Whether this form of akathisia can also lead to violence and/or suicide has not been studied.

In summary, the presence of any of these tardive movement disorders may cause a patient embarrassment as a result of the reaction of others to abnormal movements or posture. As far back as 1968, Crane (19) commented on the fact that movement disorders of a tardive type could lead to social stigmatization. In fact it has been suggested that such movement disorders are part of schizophrenia (20,21) as opposed to long-term neuroleptic toxicity and in that sense professionals and the lay public alike may associate the presence of these "visible" side effects of mental illness treatment with the mental illness itself. Clearly it is important to educate not only physicians and health care professionals about tardive movement disorders but also families and friends in order to minimize embarrassment and resultant depression.

Dysfunction

Depression in schizophrenic patients may occur as a result of a physiological dysfunction which is secondary to the presence of a tardive syndrome. Two tardive syndromes that result in physiological dysfunction are respiratory dyskinesia and tardive water intoxication.

Respiratory dyskinesia has been described as early as 1964 when Hunter and colleagues (22) commented on two patients with tardive dyskinesia who presented with "periodic disturbance of respiratory rate, rhythm and amplitude with pauses in inspiration and absence of pause between inspiration and expiration even during sleep". Other authors have described irregular respiration, at times rapid and superficial but at other times deep, in patients with tardive dyskinesia (23,24). Ayd (25) has reviewed the topic describing its characteristics which consist of irregular respiration accompanied by shortness of breath at rest and occasional audible involuntary grunting and gasping noises. The presence of respiratory dyskinesia in a patient may result in sighing movements which can cause the patient to physically resemble a depressed patient. However a form of depression can occur as a result of blood gas abnormalities secondary to respiratory dyskinesia. Respiratory alkalosis has been described in patients with respiratory dyskinesia (24). We have reported on patients developing a syndrome where blood gas abnormalities are marked during the presence of respiratory dyskinesia during the day but return to normal at night during sleep when abnormal involuntary movements cease (26). Given the potential for patients with respiratory dyskinesia to go through periods of apnea and tachypnea blood gas abnormalities can range from hypoxia to a syndrome resembling psychogenic hyperventilation. Thus in the latter the patients may resemble patients with anxiety and may in fact be improperly diagnosed and treated. In the former patients may become mildly obtruded with complaints of diminished concentration and be confused with patients suffering from mild or moderate forms of depression. In fact we have seen one patient who was treated with an antidepressant based on a suspicion of depression and subsequently developed more severe respiratory dyskinesia presumably as a result of the anticholinergic effect of the antidepressant exacerbating the dyskinesia. It is clear that the development of respiratory dyskinesia as a result of dyskinetic involvement of the diaphragm and intercostal muscles can lead to a number of complications some of which may resemble medical diagnoses and others of which may mimic psychiatric conditions such as severe anxiety and even depression. Proper diagnosis will lead to proper management. Incorrect diagnosis may lead to a number of interventions which are not necessary and potentially even harmful.

Another putative tardive syndrome associated with mood swings and depression is that of tardive water intoxication. Although self-induced water intoxication syndrome (SIWIS) has been suggested to be related to a number of causes including a number of different pharmacological agents (27-29), the mechanisms of control of water drinking behavior would seem to be related to dopamine activity in the hypothalamus. Dopaminergic control of thirst behavior has been demonstrated in several studies. Dopamine stimulates thirst (30). Haloperidol inhibits thirst when injected centrally (31,32). Long-term exposure to neuroleptics such as haloperidol has been demonstrated to result in dopamine receptor supersensitivity (30). This is one of the leading hypotheses in terms of the pathophysiology of tardive dyskinesia. The same mechanism has been proposed to result in hypothalamic dopaminergic supersensitivity which in turn could lead to increased thirst behavior (33). The result would be that long-term neuroleptic exposure would lead to dilutional hyponatremia and complications arising thereof. Clinically water intoxication has been described mainly in neurological terms. The patients develop tremor, stupor, coma and may eventually have seizures which in some cases are life threatening.

However, the syndrome of water intoxication clinically can resemble intoxication with a substance such as alcohol. Patients in addition to showing some neurological problems may become lethargic, quiet, depressed or may show disinhibition of behavior with euphoria, irritability and in combination with their illness increased psychotic symptoms (34). As a result patients may appear to be depressed if they are showing primarily stupor and lethargy as a result of water drinking but may also, through disinhibition, act impulsively with attempted suicide which again may suggest an underlying depressive illness. Furthermore the syndrome of water intoxication seems to be most apparent in the afternoon and evening hours (35). This may be the result of patients beginning to drink water after awakening and intoxicating themselves only later in the day. Thus there may appear to be a diurnal rhythm to the symptoms which may again suggest depression or mood disorder. Finally, associated with the syndrome of water intoxication is a second disorder, hypothermia. Hypothermia may result from the combination of drinking cold tap water in excess associated with a drug induced poikilothermy (36). The result is that patients become hypothermic in the afternoon and evening and may in order to warm themselves demonstrate a hunched posture, wear excessive clothing and generally withdraw emotionally. The hypothermia may also contribute to a general CNS depressive effect which can further mimic depression. Although the possibility of a tardive water intoxication syndrome mimicking depressive illness in a schizophrenic population may seem esoteric and potentially rare, it is estimated that 6-17% of chronic patients with schizophrenia demonstrate polydipsia (37). Although only a quarter to a half of polydipsia patients show clear signs of water intoxication it is possible that some have more minor manifestations of the problem which in turn might still affect mood.

Thus, through physiological dysfunction as a result of tardive syndromes such as respiratory dyskinesia and water intoxication, schizophrenic patients may demonstrate depressive-like symptoms. The management of these syndromes is very much different from the true management of depression appearing in a schizophrenic patient.

Damage

The most consistent finding from newly developed imaging techniques in terms of altered functioning of the schizophrenic brain compared to controls has been decreased activity in the frontal lobes. A preliminary report by Farkas and colleagues in never-medicated schizophrenics showed a relative metabolic hypofunction of the frontal lobe (38). This was consistent with the earlier studies of regional cerebral blood flow by Ingvar and Franzen (39). The first published controlled series with positron emission tomography (PET) compared 8 off-medication patients with schizophrenia and 6 normal controls (40). Again hypofrontality was found in the schizophrenic group. Subsequently other studies have confirmed this finding (41-45); however other studies have not (46-49). In general one can say that the finding of relative hypofrontality in schizophrenia has always been more robust in patients who have been more chronically ill and thus exposed to neuroleptics for longer periods of time. Weinberger and colleagues (50), using regional cerebral blood flow, have found reduced blood flow in the dorsolateral prefrontal cortex during performance of the Wisconsin Card Sort Test, a cognitive task designed to activate this region. They have subsequently replicated their findings in medication-free patients, however these patients had been treated with neuroleptics in the past (51,52). They point out in their discussion of their most recent replication study that neuroleptic artifact has been put forward by a number of authors as the agent responsible for hypofrontality (44-46). They deal with this argument mainly from the point of view of immediate effects of neuroleptics stating that studies of patients receiving neuroleptics have not been more likely to find hypofrontality than have studies of medication-free patients and that a number of studies have reported that neuroleptics actually increase cortical metabolism. What is not dealt with is the potential for neuroleptics to cause hypofrontality in schizophrenia through a long-term potentially irreversible toxic reaction to the frontal cortex leading to cell damage and dysfunction even after neuroleptic clearance from the brain. Is there any evidence to support this possibility? Benes and colleagues (53) studied the prefrontal cortex in schizophrenic patients who had been treated with neuroleptics chronically through examination of the cytoarchitecture of this and other regions of the brain post mortem. In their study they found decreased cell density in the prefrontal cortex compared to a normal control group. An attempt was made to control for neuroleptic exposure, however this was done by calculating the previous four weeks of treatment that patients had received despite the fact that these patients had an average age of 60. The total life time exposure to neuroleptic treatment was not taken into consideration. Thus it is not surprising that there was no correlation between neuroleptic treatment in terms of the dose received for the previous four weeks and the tendency to have cell density decreases in the schizophrenic brain, and specifically the prefrontal cortex. In fact Benes' paper refers to a study by Dunlap prior to the availability of neuroleptics which found no differences in cell densities between schizophrenic brains and normal controls. Although they point out potential artifacts in the study by Dunlap there was the obvious artifact in the study of Benes of past neuroleptic exposure. This raises the question as to whether the decreased cellular density found in neuroleptic treated schizophrenics is simply related to neuroleptic treatment and nothing else.

Szechtman and colleagues have studied prospectively a group of schizophrenic patients using PET (54). The patient groups were patients who were drug naive and then followed after one year of treatment with PET scanning on both occasions. Other patients were scanned with variable lengths of neuroleptic exposure up to 14 years. They found in the drug naive patients hyperfrontality which remained after one year of drug treatment. In contrast following 4-14 years of medication the activity pattern of chronic schizophrenics started to resemble controls. They hypothesized that greater neuroleptic exposure might even reverse the pattern of hyperfrontality yielding hypofrontality noted in chronic schizophrenic patients by others. Of interest was the corresponding finding after drug exposure that activity in the corpus striatum began to increase. Thus in never medicated schizophrenic patients, corpus stiatal activity was within the normal range. Following one year of medication, however, it had increased in all but one patient, the patient not receiving neuroleptics. The patients medicated for longer periods seemed to have even greater elevations of metabolic activity in the corpus striatum. The pattern that emerges is that as the frontal lobe activity decreases over time gradually changing from a hyperfrontal pattern to a normal pattern and even hypofrontal, corpus stiatal activity increases over time and is significantly elevated by one year and more so after at least 4 years of drug treatment. This raises the interesting possibility that neuroleptic drugs are producing a hypofrontal pattern through long-term exposure. The clinical expression of this hypofrontality could theoretically be a "frontal lobe syndrome". The frontal lobe syndrome in turn could mimic the negative or deficit symptoms of the illness but also could resemble depression.

Of associated interest is the fact that this change in ratio of activity between the frontal lobes and the corpus striatum might in some way reflect the development of dyskinesias of a tardive type. The frontal lobe has been shown in animal studies to have inhibitory influence on the caudate and the nucleus accumbens (50). One could speculate that the combination of drug-induced damage to the frontal lobe and increased activity in the striatum as a result of long-term neuroleptic exposure could lead to a clinical syndrome such as tardive dyskinesia. A similar pattern occurring in the relationship of frontal lobe activity to the nucleus accumbens might account for the clinical syndrome of tardive psychosis (55). The findings of hyperfrontality maintained after one year of drug treatment while corpus striatal activity increased after one year could reflect the patient who is developing dyskinesia but of the reversible type if neuroleptics can be discontinued. However, the patients who have been treated from 4 to 14 years and have started to show further decreases in frontal lobe activity and correspondingly more marked increases in striatal activity may demonstrate the more severe expressions of dyskinesia and other tardive syndromes in potentially irreversible forms.

CONCLUSIONS

It is proposed that long-term neuroleptic exposure may lead to depressive-like symptoms in patients with schizophrenia through a number of mechanisms. The first mechanism, disfigurement as a result of the development of tardive dyskinesia or dystonia may result in reactive depression as the patient psychologically comes to grip with not only the illness that he is suffering from but the presence of a disfiguring physical condition caused by the drugs he is receiving for treatment. Further research in this area is probably not as important as research into the development of new drugs that do not have the same propensity to cause dyskinesias and dystonias.

A second mechanism is the development of dysfunction physiologically as a result of the presence of tardive symptoms such as water intoxication or respiratory dyskinesia. Given the morbidity associated with these conditions more study is necessary to determine those at risk, the mechanism by which the syndromes arise and improved management techniques including treatments.

The third mechanism in which depressive-like symptoms may arise as a result of a tardive frontal lobe syndrome is the most speculative of the mechanisms proposed but possibly the most important. First, it is important clinically in that the neuroleptics may be causing a frontal lobe syndrome with resulting apathy, amotivation and depressive-like symptoms. Also the associated model of frontal lobe damage accompanying subcortical overactivity of dopamine systems in the corpus striatum and nucleus accumbens which might in combination result in irreversible tardive dyskinesia and possibly tardive psychosis has great clinical relevance. This combination of toxicities putatively leading to mood disorder symptoms on one hand and psychosis on the other could respectively result in schizophrenic patients resembling mood disorder patients and vice versa. Thus the diagnostic boundaries of the two functional psychoses of Kraepelin would be gradually blurred with long-term neuroleptic treatment. Secondly, beyond the clinical ramifications, the research ramifications of neuroleptic exposure leading to frontal lobe hypofrontality and defective dorsolateral prefrontal cortical activity are enormous. These latter findings if attributed to the illness of schizophrenia have ramifications for understanding the etiology of the disease and suggesting new directions in the search for new treatments. If they are, on the other hand, artifact of drug treatment then we will be travelling down a garden path. We must continue to seek out new ways to study schizophrenia so that neuroleptic artifact both during short-term administration and as a result of long-term administration can be avoided. This involves heavy emphasis on the drug naive patient as an important study candidate, the use of strategies identifying high risk individuals for study prospectively before the illness has actually appeared and indirect routes of study circumventing drug artifact such as genetic strategies. With these approaches it may be possible in the future to truly understand the nature of schizophrenia and its varied manifestations including depression. Until such strategies are adopted, old strategies using patients who are on neuroleptics or who have had extensive drug histories may result in depression not only in patients as a result of disfigurement, dysfunction or damage but will also result in "depression" of schizophrenia research through the production of confusion as to what is related to the illness and what is in fact due to the treatment.

REFERENCES

1. Alpert D, Diamond F, Friedhoff A: Tremorographic studies in tardive dyskinesia. Psychopharmacol Bull 12: 5-7, 1976.
2. Rosenbaum AH, Niven RG, Hanson NP, et al: Tardive dyskinesia: relationship with primary affective disorder. Dis Nerv Syst 38: 423-427, 1977.
3. Kane J, Struve FA, Weinhold P, et al: Strategy for the study of patients at high risk for tardive dyskinesia. Am J Psychiatry 137: 1265-1267, 1980.
4. Rush M, Diamond F, Alpert M: Depression as a risk factor in tardive dyskinesia. Biol Psychiatry 17: 387-392, 1982.
5. Hamra BJ, Nasrallah H, Clancy J, et al: Psychiatric diagnosis and risk for tardive dyskinesia. Arch Gen Psychiatry 40: 347-348, 1983.
6. Yassa R, Ghadirian AM, Schwartz G: Prevalence of tardive dyskinesia in affective disorder patients. J Clin Psychiatry 44: 410-412, 1983.
7. Yassa R, Nair V, Schwartz G: Tardive dyskinesia and the primary psychiatric diagnosis. Psychosomatics 25: 135-138, 1984.
8. Kane JM, Woerner M, Weinhold P, et al: A prospective study of tardive dyskinesia development: preliminary results. J Clin Psychopharmacol 2: 345-349, 1982.
9. Gardos G, Casey DE (eds): Tardive Dyskinesia and Affective Disorders. Washington, DC, American Psychiatric Press Inc, 1984.
10. Mukherjee S, Rosen AM, Giovanni C: Persistent tardive dyskinesia in bipolar patients. Arch Gen Psychiatry 43: 342-346, 1986.
11. Jones BD, Chouinard G, Annable L, et al: Factors related to tardive dyskinesia in affective disorders. 33rd Annual Meeting of the Canadian Psychiatric Association, No. 122, 1983.
12. Jones BD: Tardive dysmentia: further comments. Schizophr Bull 11: 187-189, 1985.
13. Freed E: Tardive dyskinesia: subjective discomfort from psychosocial stress. South African Med J 62: 80-88, 1982.

14. Weiner WJ, Goetz CG, Nausieda PA, et al: Respiratory dyskinesias: extrapyramidal dysfunction and dyspnea. Ann Intern Med 88: 327-331, 1978.
15. Chouinard G, Boisvert D, Bradwejn J: Tardive dyskinesia in a nonpsychiatric patient due to short-term use of a neuroleptic/anticholinergic combination drug. Can Med Assoc J 126: 821-822, 1982.
16. Burke RE, Fahn S, Jankovec J, et al: Tardive dystonia: late onset and persistent dystonia caused by antipsychotic drugs. Neurology 32: 1335-1346, 1982.
17. Stahl SM: Akathisia and tardive dyskinesia: changing concepts. Arch Gen Psychiatry 42: 915-917, 1985.
18. Barnes TRE, Braude WM: Akathisia variants and tardive dyskinesia. Arch Gen Psychiatry 42: 874-878, 1985.
19. Crane GE: Tardive dyskinesia in patients treated with major neuroleptics: a review of the literature. Am J Psychiatry 124: 40-48, 1968.
20. Owens DGC, Johnstone EC, Frith CD: Spontaneous involuntary disorders of movement in neuroleptic treated and untreated chronic schizophrenics: prevalence, severity, and distributions. Arch Gen Psychiatry 39: 452-461, 1982.
21. Crow TJ, Cross AJ, Johnstone EC, et al: Abnormal involuntary movements in schizophrenia: Are they related to the disease process or its treatment? Are they associated with changes in dopamine receptors? J Clin Psychopharmacol 2: 236-340, 1982.
22. Hunter R, Earl CJ, Thornicroft S: An apparently irreversible syndrome of abnormal movements following phenothiazine medication. Proc Roy Soc Med 57: 758-762, 1964.
23. Degkwitz R, Wenzel W: Persistent extrapyramidal side effects after long-term application of neuroleptics, in Neuropsychopharmacology. Edited by Brill H, Amsterdam J. International Congress Series 1967.
24. Greenberg DB, Murray GB: Hyperventilation as a variant of tardive dyskinesia. J Clin Psychiatry 42: 401-403, 1981.
25. Ayd FJ: Respiratory dyskinesias in patients with neuroleptic-induced extrapyramidal reactions. Int Drug Therapy Newsletter 14: 1-3, 1979.
26. Bassett A, Jones BD, Wilcox PG, et al: Objective evaluation of respiratory dyskinesia. 140th Meeting of the American Psychiatric Association, New Research Abstract 113, 1987.
27. Peterson DT, Marshall WH: Polydipsia and inappropriate secretion of antidiuretic hormone associated with hydrocephalus. Ann Intern Med 83: 675-676, 19875.
28. Rao KJ, Miller M, Moses A: Water intoxication and thioridazine (letter). Ann Intern Med 82: 61-63, 1975.
29. Winstead DK: Coffee consumption among psychiatric inpatients. Am J Psychiatry 133: 1447-1450, 1976.
30. Jones BD: Psychosis associated with water intoxication: psychogenic polydipsia or concomitant dopaminergic supersensitivity disorders? (letter). Lancet 2: 519-520, 1984.
31. Dourish CT: Dopaminergic involvement in the control of drinking behaviour: a brief review. Prog Neuropsychopharmacol Biol Psychiatry 7: 487-493, 1983.

32. Marshall JF, Richardson JS, Teitelbaum P: Nigrostriatal bundle damage and the lateral hypothalamic syndrome. J Comp Phys Psychol 87: 808-830, 1974.

33. Shen WW, Sata LS: Hypothalamic dopamine receptor supersensitivity? A pilot study of self-induced water intoxication. Psychiatr J Univ Ottawa 8: 154-158, 1983.

34. Arieff AI, Llach F, Massry SG: Neurological manifestations and morbidity of hyponatremia: correlation with brain water and electrolytes. Medicine 55: 121-129, 1976.

35. Koczapski AB, Ibraheem S, Ashby YT, et al: Early diagnosis of water intoxication by monitoring diurnal variations in body weight. Am J Psychiatry 144: 1626, 1987.

36. Koczapski AB, Ashby YT, Ibraheem S, et al: Afternoon radiator-sitting syndrome (ARSS): hypothermia and early diagnosis of self-induced water intoxication. Br J Psychiatry 151: 133-134, 1987.

37. Illowsky BP, Kirch DG: Polydipsia and hyponatremia in psychiatric patients. Am J Psychiatry 145: 675-683, 1988.

38. Farkas T, Reivich M, Alavi A, et al: The application of $[^{18}F]$ 2-deoxy-2-fluoro-D-glucose and positron emission tomography in the study of psychiatric conditions, in Cerebral Metabolism and Neural Function. Edited by Passonneau JV, Hawkins RA, Lust WD, Welsh FA. Baltimore, Williams & Wilkins Company, 1980.

39. Ingvar DH, Franzen G: Abnormalities of cerebral blood flow distribution in patients with chronic schizophrenia. Acta Psychiatr Scand 50: 425-462, 1974.

40. Buchsbaum MS, Ingvar DH, Kessler R, et al: Cerebral glucography with positron tomography: use in normal subjects and in patients with schizophrenia. Arch Gen Psychiatry 39: 251-259, 1982.

41. Farkas T, Wolf AP, Jaeger J, et al: Regional brain glucose metabolism in chronic schizophrenia. Arch Gen Psychiatry 41: 293-300, 1984.

42. DeLisi LE, Holcomb HH, Cohen RM, et al: Positron emission tomography in schizophrenic patients with and without neuroleptic medication. J Cereb Blood Flow Metab 5: 201-206, 1985.

43. DeLisi LE, Buchsbaum MS, Holcomb HH, et al: Clinical correlates of decreased anteroposterior metabolic gradients in positron emission tomography (PET) of schizophrenic patients. Am J Psychiatry 142: 78-81, 1985.

44. Wolkin A, Jaeger J, Brodie JD, et al: Persistence of cerebral metabolic abnormalities in chronic schizophrenia as determined by positron emission tomography. Am J Psychiatry 142: 564-571, 1985.

45. Volkow ND, Wolf AP, Van Gelder P, et al: Phenomenological correlates of metabolic activity in 18 patients with chronic schizophrenia. Am J Psychiatry 144: 151-158, 1987.

46. Volkow ND, Brodie JD, Wolf AP, et al: Brain metabolism in patients with schizophrenia before and after acute neuroleptic administration. J Neurol Neurosurg Psychiatry 49: 1199-1202, 1986.

47. Kling AS, Metter J, Riege WH, et al: Comparison of PET measurement of local brain glucose metabolism and CAT

measurement of brain atrophy in chronic schizophrenia and depression. Am J Psychiatry 143: 175-180, 1986.

48. Gur RE, Resnick SM, Alavi A, et al: Regional brain function in schizophrenia, I: a positron emission tomography study. Arch Gen Psychiatry 44: 119-125, 1987.

49. Gur RE, Resnick SM, Gur RC, et al: Regional brain function in schizophrenia, II: repeated evaluation with positron emission tomography. Arch Gen Psychiatry 44: 126-129, 1987.

50. Weinberger DR, Berman KF, See RF: Physiologic dysfunction of dorsolateral prefrontal cortex in schizophrenia, I: regional cerebral blood flow evidence. Arch Gen Psychiatry 43: 114-124, 1986.

51. Weinberger DR, Berman KF, Illowsky BP: Physiological dysfunction of dorsolateral prefrontal cortex in schizophrenia, III: a new cohort and evidence for a monoaminergic mechanism. Arch Gen Psychiatry 45: 609-615, 1988.

52. Berman KF, Illowsky BP, Weinberger DR: Physiological dysfunction of dorsolateral prefrontal cortex in schizophrenia, IV: further evidence for regional and behavioral specificity. Arch Gen Psychiatry 45: 616-622, 1988.

53. Benes FM, Davidson J, Bird ED: Quantitative cytoarchitectural studies of the cerebral cortex of schizophrenics. Arch Gen Psychiatry 43: 31-45, 1986.

54. Szechtman H, Nahmias C, Garnett S, et al: Effect of neuroleptics on altered cerebral glucose metabolism in schizophrenia. Arch Gen Psychiatry 45: 523-532, 1988.

55. Chouinard G, Jofnes BD, Annable L: Neuroleptic-induced supersensitivity psychosis. Am J Psychiatry 135: 1409-1410, 1978.

56. Chouinard G, Jones BD: Neuroleptic-induced supersensitivity psychosis: clinical and pharmacologic characteristics. Am J Psychiatry 137: 16-21, 1980.

57. Chouinard G, Jones B: Neuroleptic-induced supersensitivity psychosis: the 'Hump course' and tardive dyskinesia (letter). J Clin Psychopharmacol 2: 143-144, 1982.

ANTIDEPRESSANTS IN "DEPRESSED" SCHIZOPHRENICS

Samuel G. Siris

Hillside Hospital
Glen Oaks, New York

Schizophrenia is a disorder with many associated morbidities. Objective manifestations of psychopathology, subjective suffering, social and vocational dysfunction, and family burden are all well known and have been well described. Among the heterogeneous patternings of symptoms which may occur in conjunction with schizophrenia are those that have been labeled as "depression" because of their phenotypic resemblance to states of depression occurring in non-schizophrenic populations (1-8). Many schizophrenic patients are found to present with such a condition, involving a pessimistic, underenergized, gloomy, pleasureless, and bedraggled state. In it, their ability to concentrate may be low, their appetitive drives for food and sex may be reduced, their sleep patterns may be altered, and some may even seem to lose their will to live. Although the etiology of this sort of "depression" is not necessarily clear and may well be varied, the fact that these states are in so many ways phenocopies of cases of primary depression raises the logical question as to whether they might be responsive to treatment with antidepressant medications. This is, indeed, a very logical question because "depression" itself is a syndrome which is really a final common clinical path. Thus, even if the initiating etiology is not the same, a treatment useful for one form of depression (such as primary depression) might also benefit patients with another form of depression (such as secondary depression in schizophrenia) by interacting with points on the causal pathway distal to the primary diathesis but still proximal to the clinical expression of the "depression."

"Depression," of course is a term which can be confusing in the literature because it has been used in a number of different ways. The word depression can be used to signify an affect, a symptom, a syndrome, or a disease entity. Often the literature itself contributes to this confusion by failing to make clear in which sense it is using the term, or by switching carelessly back and forth among these meanings. Thus the literature on "depression" in schizophrenia may present ambiguities as to whether the patients involved in the discussion have a temporary lowering of their mood, have a sustained complaint with regard to a lowered mood, have a complex of symptoms including vegetative signs commonly associated with primary depression, or are having an episode of illness characterized by a biomedical diathesis leading to such a complex of symptoms. For patients with primary depressions the usefulness of antidepressant medications became more clear when the distinctions among these various uses of the term "depression" were more carefully developed. It is therefore likely that the same situation might well be the case for those forms of secondary depression which have been observed in the course of schizophrenia.

A number of situations which can present as clinical phenocopies of depression in the course of schizophrenia have been elaborated in the literature. Included among these are akinesia, akathisia, transient disappointment reactions, demoralization, negative symptoms, and post-psychotic depression. The differential diagnosis of these conditions will provide a guideline for helping to determine in whom a therapeutic trial of an antidepressant medication might be a sensible undertaking.

Akinesia is an extrapyramidal neuroleptic side effect characterized by a reduction in spontaneity (9-11). Patients with this condition have an impaired capacity to initiate activity on their own and, as a result, may spend a great deal of their time in a state of relative inactivity. Although the term akinesia was initially used to denote a specific reduction in associated motor movements, such as arm-swing while walking, as a neuroleptic-induced extrapyramidal side effect (12), and such a reduction in motor movements may make the presentation of akinesia obvious, limited associated motor movements, muscle stiffness or cogwheeling are no longer required for the modern diagnosis of akinesia. Needless to say, the syndrome of akinesia can be quite debilitating for a patient. In his or her non-spontaneous state, friends lose interest and drift away, and patients themselves are rather ineffective at being able to arrange life's requirements - let alone life's pleasures - for themselves. They find life boring and dull, and feel helpless and pessimistic about the prospect of doing anything about it. They may well not recognize that this state is a neuroleptic induced side effect, so they may blame it on themselves or assume that it is now just part of their identity. Their families may become frustrated or even infuriated with them for "not showing drive or initiative" or not seeming to care about themselves or anybody else anymore. It is therefore crucial that clinicians consider akinesia in any seemingly underenergized, undermotivated, apathetic neuroleptic-treated patient, because it is a state which can readily be reversed with reduction of neuroleptic medication or adequate treatment with antiparkinsonian agents (13-15).

Akathisia is a second neuroleptic-induced extrapyramidal side effect which can present with dysphoria as a prominent manifestation (16-17). The persistent subjective and objective manifestations of motor restlessness which constitute akathisia are generally quite unpleasant for patients experiencing them. These manifestations can come to dominate the patient's experience, interfering with pleasure, sleep, concentration, and appetitive drives. With akathisia, a patient's capacity to socialize may suffer in that they have a reduced capacity to "hold still" long enough to interact effectively or participate in social activities. As a result they may become lonely, isolated, and feel that they have "only themselves to blame." Although akathisia may sometimes respond to adjunctive antiparkinson medication, unfortunately it is often more difficult to treat than akinesia (15). On occasion, propanolol or benzodiazepines may be useful adjunctive treatments. Obviously, reduction of neuroleptic dosage is a desirable alternative approach to the treatment of akathisia, if that can be feasibly accomplished. Accurate recognition of akathisia is also important because it may be one form of dysphoric reaction in schizophrenia which might possibly be worsened by the addition of a tricyclic antidepressant to the neuroleptic regimen. This possibility was raised by a study which suggested that akathisia was one of several severe extrapyramidal side effects which seemed to result when imipramine was substituted for benztropine in a series of dysphoric schizophrenic patients (18). In that study, however, it was difficult to be certain if the extrapyramidal symptomatology was exacerbated by the introduction of imipramine, by the discontinuation of benztropine, or some specific undesirable interaction between these two events.

The next consideration in the differential diagnosis of "depression" in schizophrenia is the possibility, which has been raised in the literature, that neuroleptic medications may sometimes induce depression directly as a side effect (2, 19-28). This is a plausible hypothesis, given that neuroleptics are assumed to have their therapeutic action in schizophrenia by means of reducing dopaminergic neurotransmission, and that dopamine also has a role in the regulation of affect (29-31). The case for neuroleptic medication inducing depression, however, has largely been based on anecdotal reports which have not controlled for the possibilities of akinesia and akathisia. Moreover, longitudinal studies which have traced symptoms prospectively have shown that depression-like symptoms which appear prominent after the patient's flagrant psychoses have subsided were actually there all along during the course of the psychotic episode but tended to receive less attention in the face of the more dramatic and florid psychotic symptomatology - until the psychosis subsided and the depression-like symptoms were "revealed" (3,32). In fact, such longitudinal studies of psychotic episodes have often shown the depression-like symptomatology which coincides with the psychosis to subside during neuroleptic treatment at a time when the psychotic symptomatology is also resolving (3,32). Such an observation, of course, runs counter to the notion that depressive symptomatology is generated by neuroleptic medication. Also running counter to that notion has been the observation that the severity of depression in "depressed" schizophrenic patients shows no association with either neuroleptic dosage or neuroleptic blood level (33) and observations from the psycho-pharmacology literature that neuroleptics may have some degree of efficacy in the treatment of at least some forms of depression (34-39).

Another item on the list of differential diagnosis of depression in any patient group is a transient disappointment reaction. Reality based or symbolic disappointments naturally engender feelings of sadness. In this regard, schizophrenic patients probably go through at least as many, if not more disappointments than most people - more interpersonal rejections, more social and vocational failures, more opportunities to compare themselves unfavorably with other people. The hallmark, of course, of a transient disappointment reaction is that it is transient. It will go away with the passage of time and requires no other intervention to do so. Nonspecific support may shorten its duration or reduce its severity, but pharmacological intervention certainly is not warranted.

A more enduring disappointment reaction is the syndrome of demoralization (40). Again, schizophrenic patients may find ample issues about which to become demoralized. They realize that they have a potentially devastating, life-disrupting disorder for which there may be variably successful treatments, but for which there is no established cure. They realize that they may have to readjust their expectations of what they are likely to get out of life socially and vocationally; and they realize that they are often misunderstood and/or stigmatized. Demoralization in the face of such dilemmas certainly is not surprising. Of course it is also true that not every patient becomes demoralized so there may be biological or other vulnerability factors playing into the issue of who becomes "demoralized." Pharmacological studies, including antidepressant studies in the treatment of demoralization in schizophrenia have not been attempted, so there is no empirical data base on which to rest any theory or recommendations in this regard.

Sometimes symptoms which resemble depression occur in schizophrenic patients when they are in the process of decompensating into their next episode of psychosis (41-42). At these times patients may become withdrawn and have reduced pleasure (possibly as a mechanism for avoiding excessive stimulation). Also at these times they may develop peculiar patterns of sleeping and/or insomnia, feel weak and exhausted, concentrate poorly, and berate themselves. This potential phenocopy of depression will declare itself as psychosis over time. But if antidepressant medications are administered early on, the mistaken conclusion could be drawn that the antidepressant was somehow causally associated with the psychotic relapse that followed (43). Anecdotal reports which may have flowed from this phenomenon formed a large part of the basis for the assertion, which appeared in the older literature and has also appeared in many text books, that the use of antidepressant medications routinely leads to psychotic relapse in schizophrenia.

When the earlier literature involving the use of antidepressant medications in schizophrenia was in fact reviewed, no studies were found which specifically addressed the crucial question: Is it beneficial to add an antidepressant to the on-going neuroleptic regimen of schizophrenic patients who demonstrate specific features clinically similar to the presentation of primary depression (44)? The closest approach to this question had involved certain studies in which a post hoc retrospective analysis was undertaken of depressive features in treated schizophrenic patients (45-50). In those analyses, the finding was suggested that depressive features, notably sadness, seemed to be ameliorated when antidepressants were added to the patients' neuroleptic regiment. Since patients were not pre-selected for these features, however, these results are very difficult to interpret. In fact the whole early literature on the use of antidepressants in schizophrenia is difficult to interpret, not only because patients were not selected on the basis of any stigmata of depression, but also because patients had not had the opportunity to be diagnosed by modern operationalized criteria. If psychotic patients, some of whom may have had mania, showed average increases in measures of psychosis after the initiation of an antidepressant (in some studies with and in some studies without a concomitant neuroleptic), this does not supply a very satisfactory answer to our question about the usefulness of the addition of an antidepressant medication to the neuroleptic regimen of depressed-appearing schizophrenic patients (44).

We will therefore concentrate here, in examining the role of adjunctive antidepressants in schizophrenic patients with "post-psychotic depression," on the limited number of studies, published during the last decade or so, which have prospectively selected patients with features of depression, and then led them through a randomized double-blind trial of an antidepressant added to their on-going neuroleptic regimen. The results of these several studies have been variable, no doubt at least in part due to differences in their design. Some of these studies have appeared to indicate that the addition of the antidepressant has been helpful to these patients. Other studies have failed to develop evidence which supports that hypothesis.

DOUBLE-BLIND STUDIES ADDING AN ANTIDEPRESSANT OR PLACEBO TO A NEUROLEPTIC IN "DEPRESSED" SCHIZOPHRENIC PATIENTS:

Waehrens and Gerlach (51) reported on 17 out of 20 chronic inactive and emotionally withdrawn schizophrenic patients who completed a double-blind cross-over study in which either maprotiline or placebo was added to the patients' neuroleptic for a period of 8 weeks on each treatment. The maprotiline dose was built up to an average final dose of 138 mg/day and ratings were obtained with the Brief Psychiatric Rating Scale (BPRS) (52), the Inactivity-Withdrawal Scale (53), the Nurse Observation Scale for Inpatient Evaluation (NOSIE-30) (54), and a special scale for global preference. Little evidence was found for improvement during maprotiline treatment. The emotional withdrawal item of the BPRS favored maprotiline treatment at trend level and patients receiving maprotiline were rated as talking to others more on the activity withdrawal scale, but it is hard to give much weight to these isolated findings in the context of the number of variables examined. One transient and self-limited exacerbation of psychosis was observed in the maprotiline condition, but this occurred in a patient known to have such episodes prior to the initiation of the maprotiline. More side effects occurred during maprotiline treatment, but the three drop-outs all occurred in the placebo-added group during the first 8-week period. The authors concluded that maprotiline was without benefit in these patients.

One critique of this study is that, although patients were presented as having a clinical diagnosis of schizophrenia, no specific diagnostic criteria were given. It is also not clear to what extent the patients may have been continuing to be psychotic at the time of the trial and whether they to any further extent presented stigmata of "depression" beyond the withdrawn or anergic features which were discussed. In this study the maprotiline dose may have been as low as 50 mg/day for some patients and therefore they may not have received doses which were at a level which would have been expected to be therapeutic for patients with primary depressions. This leads to questions about the adequacy of the antidepressant trial. The cross-over design leads to some further complications. If there were any continuing effects of the maprotiline, the placebo data would be potentially compromised inasmuch as 10 out of the 17 placebo treated patients had recently finished a maprotiline treatment trial. The relatively small number of patients, of course, raises the possibility of a Type II error in the interpretation of these negative results, a possibility which is made greater by the additional variability introduced because patients were receiving a variety of neuroleptics and 60% of the patients were taking antiparkinsonian medications whereas 40% were not.

Kurland and Nagaraju (55) report on a double-blind study in which a 4 week trial of viloxazine, in doses built up to a maximum of 300 mg per day in the final week of the trial, or placebo was administered to schizophrenic patients who were stabilized on either haloperidol or chlorpromazine. In order to quality for this trial, patients were required to have a score of at least 18 on the Hamilton Depression Rating Scale (HAM-D) (56). No antiparkinsonian medications were permitted during the trial, which was completed by 22 of the 28 patients who entered. Assessments were made with the BPRS, HAM-D, Clinical Global Impression (CGI) scale (57), and Zung Self-Rating Depression Scale (58). The actual data were not presented by the authors, but they concluded as an "overall assessment" that there were "no apparent differences between the viloxazine and placebo-treated patients. Where improvement was observed, this was reflected in the ameliorization of depressive symptomatology frequently accompanied by an improvement in ward behavior."

This constitutes a difficult conclusion to evaluate in the absence of the data on which it was based. This study also suffers from the fact that no operational diagnostic criteria are given, the dose of the antidepressant may have been too low, planned dosage increases were not carried out if patients began to show substantial improvement, the duration of the treatment trial was rather short, and two quite different neuroleptics were employed which would confound the analysis of a data set which was already small and therefore prone to Type II error. It was also unclear to what extent these inpatients were or were not psychotic at the time of the trial.

Johnson (4) studied 50 chronic patients who had been diagnosed as having schizophrenia on the basis of having Feighner or Schneiderian symptoms, who also met defined criteria for depression which included a Beck Depression Inventory (BDI) (59) score of at least 15. At the time of the study, all the patients were being maintained on depot neuroleptic medication, either fluphenazine decanoate or flupenthixol decanoate. Patients were then randomized to a 5 week double-blind trial for either nortriptyline, in doses to a maximum of 150 mg per day, or placebo. Oral antiparkinsonian medications and benzodiazepines were allowed as supplemental medications. Ratings were obtained with the HAM-D instrument, the BPRS, and a side effects check list. More side effects were reported in the nortriptyline group. Although there were more individuals with good improvement of depressive symptomatology (HAM-D score reduced to less than 5) in the nortriptyline group (7 out of 25) than in the placebo group (2 out of 25), these results did not reach statistical significance and these findings were considered by the author to constitute a negative result.

This study had a more rigorously defined patient group than the previous two studies discussed, and had a larger sample of patients to study. One weakness was that the study design called for patients to enter until 25 patients in each group completed the 5 week course. It is not mentioned how many subjects, if any, dropped out of each group before that point, and if so what their scores or status was at the time. These data might affect the general interpretation of the results. The number of patients who received antiparkinson medications or benzodiazepines also is not specified. Another methodological problem with this study concerns the dose of nortriptyline which was employed. A dose as high as 150 mg/day would be expected to raise the plasma level of nortriptyline for a number of patients above the upper limit of the putative "therapeutic window" which has been described for primary depressives receiving this drug (60-63). Furthermore, the presence of concomitant neuroleptic medication would be expected to raise the tricyclic plasma level even further by competing for the hepatic pathways which metabolize the tricyclic (64-67). Of course, the therapeutic window argument has been worked out on the basis of data drawn from primary non-delusional depressives. Whether this relationship would hold for patients having schizophrenia with secondary depression or for patients being simultaneously treated with a neuroleptic has not as yet been tested. Unfortunately, in the study by Johnson, plasma tricyclic levels were not collected or reported. The possibility that at least some of the patients treated with nortriptyline may not have been in the optimal therapeutic range, however, raises the question of whether a 28% favorable response with nortriptyline versus an 8% favorable response to placebo, in terms of depression-like symptomatology, is really such a negative result with regard to the usefulness of adjunctive nortriptyline for at least some depressed schizophrenic patients.

Becker (68) studied 52 patients meeting Research Diagnostic Criteria (RDC) (69) for schizophrenia and who also met RDC for major depressive syndrome. These patients were either newly admitted or chronically hospitalized; 22 had moderate or severe subjective depression and the rest were anergic with loss of interest and impaired performance at work and activities. After a drug free period of 2 weeks (less if the patient's state deteriorated), patients were blindly randomized to either thiothixene alone (mean dose = 40 mg/day) or a combination of chlorpromazine and imipramine (mean doses of 640 and 156 mg/day, respectively) for a 4 week trial. Benztropine was used "as necessary" for extrapyramidal symptoms. Patients were rated with the BPRS, The HAM-D, and the CGI rating instruments. Both treatments were judged effective compared to baseline, but neither was found to be statistically superior to the other on any measure of psychopathology. The combination treatment group had more total side effects, but the statistical significance of this was not reported.

The results of this study are difficult to evaluate because two different neuroleptic agents were used, a low potency neuroleptic (chlorpromazine) for those patients randomized to imipramine and a high potency neuroleptic (thiothixene) for those patients randomized to placebo. Neuroleptic treatment differences may have been difficult to untangle from antidepressant/placebo differences - especially with regard to the issue of side effects. Also, the final neuroleptic doses, being given in a doctors'-choice paradigm, turned out not to be given in the 20:1 thiothixene to chlorpromazine potency ratio which the author initially anticipated, so that the patient group receiving imipramine actually received somewhat less neuroleptic. Additionally, the thiothixene patients may have gotten more antiparkinson medication, since the author noted more overt extrapyramidal side effects in that treatment group, and therefore that treatment group might have subsequently manifested less akinesia than the chlorpromazine/imipramine group - although this cannot be evaluated from the data presented. The patients involved in this study were diagnosed by stringent criteria, the RDC, but it is not clear what that diagnosis was since the author states that they met criteria both for schizophrenia and major depression. According to RDC rules, that would mean their diagnosis would be schizoaffective disorder, not schizophrenia, if these conditions were met concurrently. At any rate, it is clear that the patient sample was heterogeneous, drawn both from acute admissions and chronic patients, and drawn from one group which manifest substantial depressed mood and one group which did not. Unfortunately, the data are not broken down in terms of who manifest depressed mood and who did not, or who was psychotic at the time of the trial and who was not. Another difficulty with this trial design is that it was brief - 4 weeks - perhaps leaving insufficient time for antidepressant efficacy to be manifest. This study also differs from the other studies discussed in that neuroleptic and antidepressant were begun simultaneously. Since both groups improved, at least some patients were apparently neuroleptic responsive. It is difficult statistically to demonstrate the potential efficacy of a possible second active treatment (the antidepressant) in the presence of the new administration of a first active and efficacious treatment (the neuroleptic), due to the variance engendered by the simultaneous response to that first active treatment. This would be especially true in a heterogeneous patient population.

Chouinard et al (50) studied 96 ambulatory patients who carried hospital diagnoses of schizophrenia and who had exhibited two or more specific psychotic symptoms detailed in their report. Following two weeks of fixed dose treatment with chlorpromazine, 125 mg/day, they were randomly assigned, in a stratified design, to one of 4 treatment groups: placebo, amitriptyline 125 mg/day, perphenazine 20 mg/day, or the combination of amitriptyline and perphenazine at those doses. The treatment trial was 12 weeks and the use of antiparkinson medication was not mentioned. Evaluation was with the CGI, the BPRS, and the Inpatient Multidimensional Psychiatric Scale (IMPS) (70), as well as an evaluation for the need for supplemental chlorpromazine during the course of the trial. Few differences were found between those patients receiving either perphenazine or the combination on the one hand or amitriptyline or placebo on the other - with the perphenazine or combination treatment set of patients doing better on a number of specific measures.

This study was strong on its statistical power and on the duration of treatment. The patient population, however, did not seem to be specifically selected for any particular features of the depression syndrome. Rather they seemed to be a sample of ambulatory schizophrenic patients otherwise unselected for other psychopathological symptoms. Therefore this study, though excellent in design and execution, was not really addressing the same hypothesis as the other studies reviewed here. A question might also be raised about the dose of amitriptyline, if "depression" were to be a target in this trial, since 125 mg/day might be a less than maximally efficacious dose in patients with primary depression.

Prusoff et al (71) studied 40 ambulatory patients who met the New Haven Schizophrenia Index criteria (72) for schizophrenia and who also had depressive symptoms of sufficient intensity to receive a score of at least 7 on the Raskin Depression Scale (73). Patients were maintained on a clinically adjusted dose of perphenazine, 16-48 mg/day for at least one month before the addition of double-blind medication, and then throughout the subsequent trial. The randomized double-blind trial consisted of the addition of amitriptyline in a flexible dose, 100-200 mg/day, or matching placebo to the treatment regimen. No other psychotropic medications were permitted. Patients were evaluated after 1,2,4, and 6 months with the BPRS, HAM-D, the Raskin Depression Scale, the Symptom Checklist (SCL-90) (74), and the Social Adjustment Scale (SAS-II) (75). The data analysis was complicated by the dropout rate over a six month study. Twenty patients completed the six months out of the 35 patients who remained in the study for at least one double-blind month and who subsequently had their data counted. The strongest point in terms of findings in the analysis was at 4 months, at which point the addition of amitriptyline was significantly associated with a reduction of depressive symptomatology, but also with greater thought disorder and agitation. The overall impression, however, in terms of therapeutic outcome, favored the benefit of the combination and indicated a general improvement of many patients rather than a dramatic improvement in just a few.

Again with this study, a question can be raised about the dose of amitriptyline and whether it was the dose which might contribute to the maximal therapeutic response. The length of treatment makes this study most relevant to the long-term clinical question concerning the wisdom of combined neuroleptic/tricyclic antidepressant treatment, but also contributed to a high drop-out rate which made data analysis problematic. As with most other studies, the question of akinesia as a potential confound is unaddressed.

Singh et al (76) examined 60 chronic hospitalized patients diagnosed as having schizophrenia by the criteria of Feighner et al (77), who also had secondary depressions involving a HAM-D score of at least 18. Patients were stabilized for 4 weeks on their existing neuroleptic or neuroleptic plus antiparkinsonian regimen, had placebo added for a week, and then randomly and double-blindly either remained on placebo or had trazodone added, to a maximum dose of 300 mg/day, for a 6 week trial. Evaluations were done with the BPRS, HAM-D, NOSIE, and CGI. Statistically significant differences were found on the CGI and HAM-D (including each of the 4 HAM-D clusters) favoring the trazodone-treated group. The trazodone group also showed more improvement on the BPRS, but this fell short of statistical significance.

Patients were particularly well stabilized in this study prior to the initiation of the double-blind, and several ratings were also done prior to the double-blind "baseline". This design may have reduced variance which might be due to novelty or non-specific "placebo response" factors, making it easier for a treatment response to the antidepressant to emerge with statistical significance. The larger sample size of this study also contributed statistical power which then may have been manifested in the significance of the results. No breakdown is given with regard to which patients were and which patients were not receiving antiparkinsonian drugs, or what the dose ranges were for either the neuroleptic or antiparkinson drugs.

The final double-blind study of the addition of an antidepressant to the neuroleptic regimen of "depressed" schizophrenics is that of Siris et al (78). In this study 33 patients with diagnoses by RDC of schizophrenia or schizoaffective disorder for their most recent flagrant psychotic episode were stabilized on their clinically best-adjusted weekly doses of fluphenazine decanoate. At the point of study entry, they were either non-psychotic or only residually psychotic and each at that point met RDC criteria "in cross-section" for either major depression or minor depression, and each had a HAM-D score of at least 12, not counting the items for paranoia, derealization, and insight. Each had also maintained both their syndrome and their HAM-D score value for a minimum of 3 consecutive weekly ratings, and each had failed to have their syndrome resolve in response to a trial of benztropine 2 mg po TID which was administered in an attempt to rule out the akinesia syndrome as a confound. With the fluphenazine decanoate and benztropine doses held constant, patients were then randomized to a 6 week trial of either imipramine, built up to a maximum of 200 mg/day, or placebo. Patients were evaluated with the CGI and selected items from the Schedule for Affective Disorders and Schizophrenia (SADS) interview (79). CGI global improvement ratings favored the imipramine-treated group with statistical and clinical significance. All 4 SADS subscales for depression also favored the imipramine cohort with statistical significance, and there was no evidence for a difference between the two groups in terms of ratings for emergent psychotic symptomatology. The mean combined imipramine/desipramine plasma level in the imipramine-treated group was 261 ± 107 ng/ml.

This study is the only one to date to make a specific attempt to rule out the possible confound of akinesia with a preliminary antiparkinsonian trial in all patients. This study is also the only study involving the treatment of depressed schizophrenic patients to measure antidepressant plasma levels. Although no relationship was found in this limited series between the antidepressant plasma levels and clinical response, the levels were indeed in the range which has generally been found to be efficacious in primary unipolar non-delusional depressed patients (80,81). The slow rate at which the imipramine dose was raised (50 mg/day per week) and the relatively brief treatment course (6 weeks) makes it unclear if the full imipramine effect was evaluated in this study (82), or what the implications might be for more prolonged treatment courses (83,84).

Fewer studies have examined the question of the use of MAO inhibitors in schizophrenia than have examined the use of tricyclic antidepressants. In fact, no prospective double-blind study has prospectively targeted the use of an MAO inhibitor to schizophrenic patients who present with specific depression-like symptoms, which would be the proper test of the question addressed here. One study, at the point of data analysis, showed with statistical significance that "depression" was ameliorated when tranylcypromine was added to trifluoperazine, (49) but it is quite open to question if the patients involved (diagnosed at the time as pseudoneurotic schizophrenia) would be diagnosed as having schizophrenia by current operationalized diagnostic criteria. In general, studies of MAO inhibitors, either alone (44,85) or in combination with neuroleptics (44,86) in rather unselected series of "schizophrenic" patients tended not to be involved with any significant improvement in double-blind studies. What may also be worthy of note, however, is that significant deterioration was not observed either. Since these studies were drawn from an earlier psychopharmacology literature before the current operationalized diagnostic schemes had been devised, and since no attempt had been made to preselect patients with specific depression-like symptoms or syndromes, the issue of the potential usefulness of MAO inhibitors in depressed schizophrenics must be described as unresolved.

In summary, there have been only a limited number of prospective double-blind controlled trails of the addition of supplemental antidepressant to the neuroleptic treatment regimen of schizophrenic patients who were selected on the basis of symptoms resembling the clinical presentation of depression at the time of the trial. In these studies, the results have been mixed. However, the studies which failed to find a positive effect, by and large, appear to have more methodological flaws than the studies which did find a benefit to the adjunctive antidepressant. Notably, most trials failed to make an attempt to eliminate the confound of neuroleptic induced akinesia, which can manifest close clinical phenocopy to depressive-like symptomatology (11). Also, in a number of the negative trials, the dosage and duration of the antidepressant treatment were suspect. The results are inconclusive with regard to whether or not the addition of an antidepressant predisposes to the exacerbation of psychotic symptomatology. This issue may involve the extent to which patients are still acutely psychotic at the point when the antidepressant trial is initiated, how well "covered" the patient is with neuroleptic medication at the time of the antidepressant trial, or how abruptly the dose of antidepressant is raised in the course of the trial.

RECOMMENDATIONS

Recommendations for when and how to use antidepressant medications in schizophrenic patients with concomitant features of depression cannot as yet be considered to be firmly established. Considerably more research needs to be accomplished to define who should receive an antidepressant, which dose of which antidepressant, for how long, and in combination with which other medications. Nevertheless, certain tentative guidelines can now be drawn which represent the current state of the art.

When a patient with schizophrenia develops depression-like symptomatology, the first step should not be immediately to introduce an adjunctive antidepressant. Instead, the first step is that the patient should be observed carefully and psychosocial supports reinforced. If the new symptomatology is a harbinger of an incipient psychotic decompensation, that can then be rapidly intercepted and appropriately and rapidly treated with additional neuroleptic medication - hopefully avoiding a hospitalization and the psychosocial insult of the episode. If on the other hand, the new symptomatology is a transient disappointment reaction, it will resolve on its own and require little further attention. If, however, psychotic symptoms do not supervene but the depression-like picture persists, further attention to the status of the patient's psychopharmacological management is indicated. The first question, in this case, is if the picture represents a neuroleptic side effect, extrapyramidal or otherwise. If possible, then, an attempt should be made to lower the patient's neuroleptic dosage. Although the patient should be watched for early signs of decompensation, patients may also begin to do much better from a psychosocial perspective with the lowering of neuroleptics and the clinician ought not to overreact to minor manifestations of psychosis which are within the patient's capacity to understand and tolerate. Psychoeducational interventions may be very helpful to patients in allowing them to remain well compensated despite occasional non-disruptive psychotic symptomatology. If the neuroleptic cannot be lowered any further, though, because aspects of psychosis become intolerable or are causing the patient to be more dysfunctional, antiparkinsonian medication needs to be vigorously pushed. The point merits emphasis that the manifestations of neuroleptic-induced akinesia may be subtle or insidious, at times occurring without simultaneous stiffness and rigidity. Only a thorough trial of antiparkinson medication will determine if a patient might respond. Since the anticholinergic antiparkinsonian medications are so variable in their metabolism, it is reasonable to keep raising the dose until either improvement occurs or anticholinergic side effects (such as dry mouth, constipation, blurry vision, etc.) appear. If subtle akathisia is suspected, and the patient has not responded to antiparkinsonian medication, a trial of a beta blocker or benzodiazepine may be attempted. At any rate, treatment with an antidepressant should not be initiated until the clinician is satisfied that the patient is on his or her optimum dose of neuroleptic and does not respond to the addition of an antiparkinsonian agent.

When an antidepressant is added, there is evidence that it may be best to continue the antiparkinsonian medication as well as the neuroleptic, bearing in mind that anticholinergic side effects appear, gradual reduction in antiparkinson medication would be indicated until the problem is resolved. The proper studies have not been done to compare different antidepressants in this situation, but there is more empirical support for the use of a tricyclic-type antidepressant than for a MAO inhibitor. When using the antidepressant, it may be best to increase the dose gradually, but to continue upward to full doses which would be considered therapeutic in cases of primary depression. For drugs which have commercially available blood level testing, such testing is advisable not only for the issue of assuring compliance, but also for documenting that excessive antidepressant blood levels do not occur - especially since neuroleptics and tricyclic antidepressants may interfere with each other's metabolism. Adjunctive antidepressant trials should be continued for adequate durations. In fact, there is limited evidence that schizophrenic patients with depression may take longer than patients with primary depression to respond to the addition of antidepressant medication. Little is known about the advisability of a trial of a second adjunctive antidepressant in schizophrenia if the first antidepressant trial did not work. The addition of further adjuncts such as lithium, along with the antidepressant, is another possibility, undocumented in the literature for schizophrenia, which may be considered since it is sometimes helpful in unresponsive cases of primary depression. In any case, it is probably indicated to maintain neuroleptic medication during the course of any antidepressant trial.

Little is known about continuation and/or maintenance treatment with supplemental antidepressant medication in those schizophrenic patients who have responded favorably. This issue has been examined in only one short series (83,84). In that cohort the indication is that longer term maintenance treatment with the adjunctive antidepressant may be beneficial and that, fortunately, such maintenance does not seem to be associated with an increased incidence of psychotic relapse.

ACKNOWLEDGEMENT

This work was supported, in part, by grants #MH34309 and DA05039.

REFERENCES

1. McGlashan TH, Carpenter WT Jr.: Post-psychotic depression in schizophrenia. Arch Gen Psychiatry 33: 231-241, 1976.
2. Falloon I, Watt DC, Shepherd M: A comparative controlled trial of pimozide and fluphenazine decanoate in the continuation therapy of schizophrenia. Psychol Med 8: 59-70, 1978.
3. Knights A, Hirsch SR: 'Revealed' depression and drug treatment for schizophrenia. Arch Gen Psychiatry 38: 806-811, 1981.
4. Johnson DAW: Studies of depressive symptoms in schizophrenia. Br J Psychiatry 139: 89-101, 1981.
5. Siris SG, Harmon GK, Endicott J: Post-psychotic depressive symptoms in hospitalized schizophrenic patients. Arch Gen Psychiatry 38: 1122-1123, 1981.
6. Mandel MR, Severe JB, Schooler NR, et al: Development and prediction of post-psychotic depression in neuroleptic-treatment schizophrenics. Arch Gen Psychiatry 39: 197-203, 1982.
7. Roy A, Thompson R, Kennedy S: Depression in chronic schizophrenia. Br J Psychiatry 142:465-470, 1983.
8. Martin RL, Cloninger CR, Guze SB, et al: Frequency and differential diagnosis of depressive syndrome in schizophrenia. J Clin Psychiatry 46(11 Sec 2): 9-13, 1985.
9. Rifkin A, Qitkin F, Klein DF: Akinesia: A poorly recognized drug-induced extrapyramidal behavioral disorder. Arch Gen Psychiatry 32: 672-674, 1975.
10. Van Putten T, May PR: "Akinetic depression" in schizophrenia. Arch Gen Psychiatry 35: 1101-1107, 1978.
11. Siris SG: Akinesia and post-psychotic depression: A difficult differential diagnosis. J Clin Psychiatry 48: 240-243, 1987.
12. Chien C, DiMascio A, Cole JO: Antiparkinson agents and depot phenothiazine. Am J Psychiatry 131: 86-90, 1974.
13. Rifkin A, Quitkin F, Kane J, et al: Are prophylactic antiparkinson drugs necessary? Arch Gen Psychiatry 35:483-489, 1978.
14. Van Putten T: Adverse psychological (or behavioral) responses to antipsychotic drug treatment of schizophrenia, in Schizophrenia and Affective Disorders: Biology and Drug Treatment edited by Rifkin A. Boston, John Wright/PSG, 1983.
15. Siris SG: Pharmacological treatment of depression in schizophrenia, in Depression and Schizophrenia. Edited by Delisi, L. Washington, American Psychological Association, in press.
16. Van Putten T: The many faces of akathisia. Compr Psychiatry 16: 43-47, 1975.
17. Siris SG: Three cases of akathisia and "acting out." J Clin Psychiatry 46: 395-397, 1985.

18. Siris SG, Rifkin A, Reardon GT, et al: Comparative side effects to imipramine, benztropine, or their combination in patients receiving fluphenazine decanoate. Am J Psychiatry 140: 1069-1071, 1983.
19. deAlarcon R, Carney MWP: Severe depressive mood changes following slow-release intramuscular fluphenazine injection. Br Med J 3: 564-567, 1969.
20. Johnson DAW, Malik NA: A double-blind comparison of fluphenazine decanoate and flupenthixol decanoate in the treatment of acute schizophrenia. Acta Psychiatr Scand 51: 257-267, 1975.
21. Ayd F: The depot fluphenazines: A reapprisal after 10 years' clinical experience. Am J Psychiatry 132: 491-500, 1975.
22. Floru L, Heinrich K, Wittek F: The problem of post-psychotic schizophrenic depressions and their pharmacological induction. Int Pharmacopsychiat 10: 230-239, 1975.
23. Singh MM, Kay SR: Dysphoric response to neuroleptic-treatment in schizophrenia: Its relationship to autonomic arousal and prognosis. Biol Psychiatry 14: 277-294, 1979.
24. Hogarty GE, Schooler N, Ulrich R, et al: Fluphenazine and social therapy in the aftercare of schizophrenic patients. Arch Gen Psychiatry 36: 1283-1294, 1979.
25. Ananth J, Chadirian A: Drug-induced mood disorder. Int Pharmacopsychiat 15: 58-73, 1980.
26. Galdi J, Rieder RO, Silber D, et al: Genetic factors in the response to neuroleptics in schizophrenia: A pharmacogenetic study. Psychol Med 11: 713-728, 1981.
27. Galdi J: The causality of depression in schizophrenia. Br J Psychiatry 142: 621-625, 1983.
28. Van Putten T, Marder SR: Low dose treatment strategies. J Clin Psychiatry 47(5) Suppl: 12-16, 1986.
29. Gerner RH, Post RM, Bunney WE Jr.: A dopamine mechanism in mania. Am J Psychiatry 133: 1177-1180, 1976.
30. Post RM: Biochemical changes during the switch processes. In Bunney WE Jr (moderator): The switch process in manic-depressive psychosis. Ann Intern Med 87: 319-335, 1977.
31. Wise RA: Brain dopamine and reward, in Progress in Psychopharmacology. Edited by Cooper SJ. New York, Academic Press, 1981.
32. Hirsch SR: Depression 'revealed' in schizophrenia. Br J Psychiatry 140: 421-424, 1982.
33. Siris SG, Sellew AP, Frechen K, et al: Antidepressants in the treatment of post-psychotic depression in schizophrenia: Drug interactions and other considerations. J Clin Chem 34: 837-840, 1988.
34. Klein DF: Importance of psychiatric diagnosis in prediction of clinical drug effects. Arch Gen Psychiatry 16: 118-126, 1967.
35. Nelson JC, Bowers MB: Delusional unipolar depression: Description and drug response. Arch Gen Psychiatry 35: 1321-1328, 1978.
36. Klein DF, Gittleman R, Quitkin F, Rifkin A: Diagnosis and Drug Treatment of Psychiatric Disorders: Adults and Children, Second Edition. Baltimore, Williams & Wilkins, 1980.

37. Charney DS, Nelson JC: Delusional and non-delusional unipolar depression: Further evidence for distinct subtypes. Am J Psychiatry 138: 328-333, 1981.
38. Brown RP, Francis A, Koscis JH, et al: Psychotic vs. non-psychotic depression: Comparison of treatment response. J Nerv Ment Dis 170: 635-637, 1982.
39. Spiker DG, Weiss JC, Dealy RS, et al: The pharmacological treatment of delusional depression. Am J Psychiatry 142: 430-436, 1985.
40. Frank JD: Persuasion and Healing. Baltimore, Johns Hopkins University Press, 1973.
41. Docherty JP, vanKammen DP, Siris SG, et al: Stages of onset of acute schizophrenic psychosis. Am J Psychiatry 135: 420-426, 1978.
42. Herz MI, Melville C: Relapse in schizophrenia. Am J Psychiatry 137: 801-805, 1980.
43. Siris SG, Rifkin A, Reardon GT, et al: Stability of the post-psychotic depression syndrome. J Clin Psychiatry 47: 86-88, 1986.
44. Siris SG, vanKammen DP, Docherty JP: The use of antidepressant medication in schizophrenia: A review of the literature. Arch Gen Psychiatry 35: 1368-1377, 1978.
45. Michaux MH, Kurland AA, Agallianos DD: Chlorpromazine-chlordiazepoxide and chlorpromazine-imipramine treatment of newly hospitalized, acutely ill psychiatric patients. Curr Ther Res 8: 117-152, 1966.
46. Hanlon TE, Ota KY, Kurland AA: Comparative effects of fluphenazine, fluphenazine-chlordiazepoxide, and fluphenazine-imipramine. Dis Nerv Syst 31: 171-177, 1970.
47. Pishkin V: Concept identification and psychophysiological parameters in depressed schizophrenics as functions of imipramine and nialamide. J Clin Psychol 28: 335-339, 1972.
48. Kurland AA, Hanlon TE, Ota KY: Combinations of psychotherapeutic drugs in the treatment of the acutely disturbed psychiatric patient, in Advances in Neuropsychopharmacology. Edited by Vinar O, Votava Z, Bradley PB. Amsterdam, North-Holland Publishing Co, 1971.
49. Hedberg DL, Houck JH, Glueck BC Jr.: Tranylcypromine-trifluoperazine combination in the treatment of schizophrenia. Am J Psychiatry 127: 1141-1146, 1971.
50. Chouinard G, Annable L, Serrano M, et al: Amitriptyline-perphenazine interaction in ambulatory schizophrenic patients. Arch Gen Psychiatry 32: 1295-1307, 1975.
51. Waehrens J, Gerlach J: Antidepressant drugs in anergic schizophrenia: A double-blind cross-over study with maprotiline and placebo. Acta Psychiatr Scand 61: 438-444, 1980.
52. Overall J, Gorham D: The brief psychiatric rating scale. Psychol Rep 10: 799-812, 1962.
53. Venables P: A short scale for rating "activity-withdrawal" in schizophrenia. J Ment Sci 103: 197-199, 1957.
54. Honigfeld G, Klett SJ: The nurses observation scale for inpatient evaluation. A new scale for measuring improvement in chronic schizophrenia. J Clin Psychol 21: 65-71, 1965.

55. Kurland AA, Nagaraju A: Viloxazine and the depressed schizophrenic - Methodological issues. J Clin Pharmacol 21: 37-41, 1981.
56. Hamilton M: A rating scale for depression. J Neurol Neurosurg Psychiatry 23: 56-62, 1960.
57. Guy W: ECDEU Assessment Manual, CDHEW Publication No. 76-338, 1976.
58. Zung WWK: A self-rating depression scale. Arch Gen Psychiatry 12: 63-70, 1965.
59. Beck AT, Ward CH, Mendelson M, et al: An inventory for measuring depression. Arch Gen Psychiatry 4: 561-571, 1961.
60. Asberg M, Cronholm B, Sjoqvist F, et al: Relationship between plasma level and therapeutic effect of nortriptyline. Br Med J 3: 331-334, 1971.
61. Kragh-Sorensen P, Asberg M, Hansen C: Plasma nortriptyline levels in endogenous depression. Lancet 1: 113-115, 1973.
62. Kragh-Sorensen P, Hansen C, Baastrup P, et al: Self-inhibiting action of nortriptyline's antidepressive effect at high plasma levels. Psychopharmacologia 45: 305-312, 1976.
63. Zeigler V, Clayton P, Taylor J, et al: Nortriptyline plasma levels and therapeutic response. Clin Pharmacol Ther 20: 458-563, 1976.
64. Vandel B, Vandel S, Allers G, et al: Interaction between amitriptyline and phenothiazine in man: Effect on plasma concentration of amitriptyline and its metabolite nortriptyline and the correlation with clinical response. Psychopharmacology (Berlin) 65: 187-190, 1979.
65. Kragh-Sorensen P, Borga G, Carle L, et al: Effect of simultaneous treatment with low doses of perphenazine on plasma and urine concentrations of nortriptyline and 10-hydroxynortriptyline. Eur J Clin Pharmacol 11: 479-483, 1977.
66. Nelson JC, Jatlow PI: Neuroleptic effect on desipramine steady state plasma concentrations. Am J Psychiatry 137: 1232-1234, 1980.
67. Siris SG, Cooper TB, Rifkin A, et al: Plasma imipramine concentrations in patients receiving concomitant fluphenazine decanoate. Am J Psychiatry 139: 104-106, 1982.
68. Becker RE: Implications of the efficacy of thiothixene and a chlorpromazine-imipramine combination for depression in schizophrenia. Am J Psychiatry 140: 208-211, 1983.
69. Spitzer RL, Endicott J, Robins E: Research diagnostic criteria: Rationale and reliability. Arch Gen Psychiatry 35: 773-782, 1978.
70. Lorr M, Klett CJ: Inpatient Multidimensional Psychiatric Scale. Palo Alto, Calif, Consulting Psychologists Press, 1966.
71. Prusoff BA, Williams DH, Weissman MM, et al: Treatment of secondary depression in schizophrenia. Arch Gen Psychiatry 36: 569-575, 1979.
72. Astrachan BM, Harrow M, Adler D, et al: A checklist for the diagnosis of schizophrenia. Br J Psychiatry 121: 529-539, 1972.

73. Raskin A, Schuterbrandt JG, Reading N, et al: Differential response to chlorpromazine, imipramine, and placebo. Arch Gen Psychiatry 23: 164-173, 1970.
74. Derogatis LR, Lipman R, Covi L: SCL 90. An outpatient psychiatric scale. Preliminary report. Psychopharmacol Bull 9: 13-27, 1973.
75. Schooler N, Hogarty J, Weissman MM: in Resource Materials for Community Mental Health Program Evaluations, Edited by Hargreaves WA, Attkisson CC, Sorensen JE. Second edition, US Dept of Health, Education, and Welfare publication. ADM 77328, 1977.
76. Singh AN, Saxena B, Nelson HL: A controlled clinical study of trazodone in chronic schizophrenic patients with pronounced depressive symptomatology. Curr Ther Res 23: 485-501, 1978.
77. Feighner JP, Robins E, Guze SG, et al: Diagnostic criteria for use in psychiatric research. Arch Gen Psychiatry 26: 57-63, 1972.
78. Siris SG, Morgan V, Fagerstrom R, et al: Adjunctive imipramine in the treatment of post-psychotic depression: A controlled trial. Arch Gen Psychiatry 44: 533-539, 1987.
79. Endicott J, Spitzer RL: A diagnostic interview: The schedule for affective disorders and schizophrenia. Arch Gen Psychiatry 35: 837-844, 1978.
80. Gram I, Reisby N, Ibsen I, et al: Plasma levels and antidepressant effect of imipramine. Clin Pharmacol Ther 19: 318-324, 1976.
81. Glassman AH, Perel JM, Shostak M, et al: Clinical implications of imipramine plasma levels for depressive illness. Arch Gen Psychiatry 34: 197-204, 1977.
82. Siris SG, Adan F, Cohen M, et al: Targeted treatment of depression-like symptoms in schizophrenia. Psychopharmacol Bull 23: 85-89, 1987.
83. Siris SG, Strahan A: Continuation and maintenance treatment trials of adjunctive imipramine in post-psychotic depression. J Clin Psychiatry, in press.
84. Siris SG, Cutler J, Owen K, et al: Maintenance imipramine treatment following post-psychotic depression. Psychopharmacol Bull, in press.
85. Brenner R, Shopsin B: The use of monoamine oxidase inhibitors in schizophrenia. Biol Psychiatry 15: 633-647, 1980.
86. Siris SG, Rifkin A: Drug treatment of depression in patients with schizophrenia, in Schizophrenia and Affective Disorders: Biology and Drug Treatment. Edited by Rifkin A. Boston, John Wright PSG, 1983.

THE NEUROLEPTIC DYSPHORIA SYNDROME

Theodore Van Putten, Stephen R. Marder
and Nicole Chabert

Veterans Administration Medical Center,
Psychiatry Division, Los Angeles, California

INTRODUCTION

In our experience, higher doses of neuroleptics are often associated with a dysphoria syndrome. We believe that this dysphoria is underrecognized, and that it is readily correctable by decreasing the dose. Naturalistic studies which compare depression ratings in patients on different doses of neuroleptics are not likely to support such a concept since dysphoric patients either leave treatment or tend to tolerate only a low dose. To study dose-related dysphoria, therefore, patients need to be randomly assigned to fixed doses of antipsychotic drug.

Dose Comparison Studies

Controlled dosage comparison studies in chronically hospitalized, treatment-refractory schizophrenics (1-13) indicate, on balance, that higher doses do not result in increased depression. These results, however, may not generalize to the usual newly admitted or readmitted schizophrenic. For one thing, the patient's drug absorption, metabolism, and tissue levels -- and so his or her response -- may be different. For another, depression is difficult to diagnose in the chronically ill with their blunted affect. Further, side effects associated with high dose might seem more depressing to a person who is expected to live and make it on his own outside of the hospital than for an institutionalized patient.

Dose Comparison Studies with Haloperidol

Since its introduction into the United States in 1967, the dosage of haloperidol has increased, even though there is no indication that higher doses are more effective (14,15). A typical dose of haloperidol (HPL) for a psychotic inpatient in 1984 was 25 mg daily (16) and, with few exceptions (17), has probably remained the same. Surprisingly few dose-comparison studies have been done, and most were for brief intervals, addressing the question of "rapid neuroleptization." (18-20).

To our knowledge, there are only two dose-comparison studies with haloperidol in the usual, newly (re)admitted schizophrenic patient (21,22) and in only one (22) were patients randomly assigned to fixed doses. Nevertheless, these two studies claim that haloperidol was well tolerated in the 10-90 mg daily dosage range. In other words, these studies support the notion (and current practice) that haloperidol has a wide therapeutic index.

Modestin et al (21), in a double-blind design compared 5 vs. 15 mg haloperidol tablets in newly admitted "acutely decompensated" schizophrenic patients. The number of tablets did not exceed six a day but could be varied accordingly to the condition of each patient. On the average, patients in the 5 mg tablet group averaged 20 mg of haloperidol (4.0 tablets) a day, and patients in the 15 mg tablet group averaged 58.0 mg haloperidol (3.9 tablets) a day. A significant amelioration of psychopathology as measured by the BPRS was observed in both groups at all rating periods (day 3, 7, 14, and 21). Patients in the low dose group improved on somatic concern and anxiety, whereas the high dose group did not, but there was no significant difference between the two groups. An EPS rating scale was used, and there were no statistically significant differences between the groups. Excluding administrative reasons, 13 patients dropped out because of poor response: 5 in the low- and 8 in the high dose group. The authors mention that "acute dyskinesia" (motoric manifestation of akathisia?) was observed in 8 drop-outs; 7 in the high dose group (suggesting perhaps that some of the high-dose drop-outs may have dropped out because of drug-related dysphoria or severe akathisia). The authors, however, concluded that "there were no differences with regard to therapeutic efficacy or tolerance to the treatment," a conclusion which, in our opinion, is not warranted if only because patients were not assigned to a fixed dose.

Donlon et al (22) randomly assigned 63 acutely ill schizophrenic patients to three oral haloperidol dosage regimens for a 10 day study period. The "low" dose group received 10 mg of haloperidol a day for the 10 days. The "high" dose group received 20 mg on the first day, and dosage was increased by 20 mg each day until there was either a remission of symptoms, limiting side effects developed, or the maximum dose of 100 mg of haloperidol was reached. The "moderate" dose group began on 10 mg of haloperidol, and dosage was increased by 10 mg each day until patients received either haloperidol 100 mg on day 10, or patients improved or developed limiting side effects as above. Each patient received ten capsules each day such that patients in the low dose group received one 10 mg haloperidol capsule and 9 placebo capsules. Using this dosage regimen, the mean highest study dosage in the "high" dose group was 90 mg and in the "moderate" group 77 mg. Again, the BPRS and the CGI were the outcome measures, and there were no differences among the three dosage groups by analysis of covariance. High dosage patients experienced more drowsiness, and the high and moderate dosage patients more dystonia (65% of the high and moderate dose groups experienced dystonia vs. 45% of the low dose group) but side effect differences were not statistically significant. An EPS rating scale was not used.

The focus in the aforementioned studies was clearly on antipsychotic effect. Side effects (particularly dysphoric emotions) were a secondary concern (only 2 out of the 5 aforementioned studies had an EPS rating scale). It would also appear that there was little personal involvement between the raters and their patients. Nevertheless, one could reasonably conclude from these studies that higher doses of haloperidol (up to 90 mg daily), at least for the first week or two, are effective, well-tolerated and pose no special risk. Our current findings suggest otherwise.

SUBJECTS AND METHODS

Eighty newly (re)admitted drug-free (for at least two weeks, but usually several months) schizophrenic (by DSM III) men were randomly assigned to receive haloperidol either 5, 10 or 20 mg daily for four weeks. In case of non-response, the doctor could increase (or decrease) the dose for another four weeks according to his clinical judgment. Clinical response was measured at baseline and weekly for the first four weeks, and at week 8 after the flexible dose period.

Briefly, these patients were in their early thirties, and had four previous hospitalizations on average. They had all served in the Armed Forces, had worked an average of 4.5 years at some time, were currently judged "markedly ill" on a nurses' rating scale, and scored at least "Moderate" on conceptual disorganization, unusual thought content or hallucinatory behavior on the Brief Psychiatric Rating Scale (PBRS) (23). In fact, their mean baseline BPRS-Schizophrenia factor score was 13 (normal = 3; maximum = 21) indicating they were quite psychotic at baseline.

RESULTS AND DISCUSSION

One of the most important findings of this study was that a 20 mg daily oral dose of haloperidol apparently had "psychotoxic" side effects, particularly after the first week of treatment. Specifically, patients assigned to the 20 mg dose left the hospital against medical advice significantly more often (35% vs. 4% for 20 mg vs. the 5 and 10 mg). They also experienced a substantial, clinically significant worsening with respect to BPRS-Withdrawal Retardation (blunted affect, emotional withdrawal, motor retardation) and akinesia by the second week of treatment. These findings have to be balanced against an earlier improvement in psychotic symptomatology during the first week (as judged by the CGI) and a tendency toward superior antipsychotic effect throughout in those who could tolerate it. For the first week of treatment 20 mg of haloperidol appeared to be the best dose. After seven days of treatment, the success rate (proportion of patients who remained in the study "much improved" without dropping out) for the 5, 10 and 20 mg doses were, respectively, 6%, 33% and 47% (Chi Square = 7.54; p = .02).

These data also show that 5 mg of haloperidol was adequate for some of these newly admitted, severely psychotic in-patients. For patients who tolerate neuroleptics poorly, usually because of extrapyramidal side effects, about 5 mg of haloperidol daily may be a realistic dose.

Higher drop-out rates at higher doses have been noted in out-patient dose-comparison studies with fluphenazine decanoate (24,25) suggesting that our findings apply to other populations. A risk of excessive or excessively prolonged high doses may be non-compliant or uncooperative behavior as well as greater iatrogenic distress, probably due to akathisia. Since patients with a history of severe EPS were either excluded or given prophylactic benztropine, it is likely that the drop-out rate would be even higher in a broader sample of newly-admitted patients.

The higher akinesia ratings on the 20 mg dose of haloperidol are also important. First, the low initiative and spontaneity (psychic and motoric) that characterize mild akinesia may adversely affect social adjustment (26,27). Second, the increased emotional withdrawal and blunted affect at the 20 mg dose of haloperidol in this study seemed to be primarily a manifestation of akinesia (r akinesia vs. BPRS-Withdrawal Retardation = .43, p = .0001; pooled within-subject correlations).

More important, akinesia may seem to produce a type of "improvement." The patient may talk less of psychotic concerns, but also talks less altogether; he seems to be less bothered by hallucinations, but he is less reactive to all else as well. Many patients with akinesia experience a peculiar absence of emotions, appear emotionally flat and often, when asked, reply that everything is alright. This is a type of "improvement" that is acceptable when a patient is tormented by intractable psychotic experiences. In such cases, an akinetic dampening down of the entire mental life may indeed be necessary and "therapeutic." (see Table I for correlations between akinesia and BPRS-Schizophrenia factor).

Kane et al (28), in a maintenance study which compared fluphenazine decanoate 1.25-5 vs. 12.5-50 mg every two weeks also found statistically significant elevations in blunted affect, emotional withdrawal, tension and motor retardation at the higher dose. He, too, proposed that subtle drug induced neurologic side effects (akinesia) accounted for those dosage effects.

We regard akathisia as the most distressing extrapyramidal side effect. Aside from the discomfort of akathisia itself, akathisia is significantly related to the BPRS-Depression factor, within the individual patient (r = .32); p = .0001) and also across patients (see Table 1). There are also substantial correlations between akathisia and the subjective SCL-90 ratings (the six correlations between akathisia and the global distress scores on the SCL-90 at weeks 2, 3 and 4 were significant, ranging from 0.36 (p = .07) to 0.61 (p = .003). Further, drop-outs at all time

Table 1   Correlation Coefficients ($\underline{r}$) between Akathisia and
           BPRS-Depression Factor (BPRS-D), Akinesia and PBRS
           -Withdrawal Retardation Factor (BPRS-R), Akinesia
           and BPRS-Schizophrenia Factor (BPRS-S)

Week	1	2	3	4	8
n	57	57	53	41	14
Akathisia/BPRS-D	.43	.46	.26	.31	.60
p	.0005	.0003	.06	.06	.02
Anxiety	.40	.42	.22	.18	.55
p	.0009	.0005	n.s.	n.s.	.04
Guilt	.33	.19	.12	.13	.45
p	.007	n.s.	n.s.	n.s.	.1
Depression	.42	.43	.19	.20	.60
p	.0005	.0004	n.s.	.16	.02
Akinesia/BPRS-R	.21	.38	.33	.48	.66
Emotional/Withdrawal	.03	.09	.03	.30	.47
p	n.s.	n.s.	n.s.	.04	.09
Motor Retardation	.60	.75	.76	.81	.97
p	.0001	.0001	.0001	.0001	.0001
Blunted Affect	.18	.12	.13	.44	.35
p	n.s.	n.s.	n.s.	.002	.2
Akinesia/BPRS-S	-.17	-.26	-.30	-.48	-.41
p	n.s.	.05	.03	.004	.07

intervals had higher akathisia ratings.   Lastly, in a prior
study of outpatients at this Center (24), we also found
significant relationship between akathisia and both
depression and anxiety.   For these reasons we regard
akathisia as a substantial contributor to dysphoric states.
There is even a case report literature linking neuroleptic-
induced akathisia and suicide (29-31) and violence (32-34).

Table 1 shows the relationships between akathisia and
Depression, akainesia and Withdrawal-Retardation; akinesia
and Schizophrenia factors on the BPRS over time.   Over time,
the relationship between akinesia and BPRS-Withdrawal-
Retardation factor became stronger reflecting the rise in
akinesia, particularly on the 20 mg dose.   The relationship
between akathisia and Depression decreased during the first
four weeks reflecting the excess dropout of patients with
higher akathisia scores, the use of higher doses of
benztropine to control severe akathisia and, when this was
not effective, decreasing the dose.

Table 1 also shows a rise in the correlation between
akathisia and BPRS-Depression and between akinesia and BPRS-
Withdrawal-Retardation between weeks 4 and 8.   This rise
reflects an increase both in akathisia and akinesia scores,
as the dose of haloperidol was increased in the 14 patients
who responded poorly during the first four weeks of
treatment.   As akathisia increased, so did BPRS-rated
Anxiety, Depression and Guilt; as akinesia increased, so did

BPRS-rated Blunted Affect, Emotional Withdrawal and Motor Retardation.

The mean increase in the daily dose of haloperidol in the 14 non-responders was modest (from 9.3 + 3.9 mg to 19.6 + 9.3 mg). Nevertheless, this modest increase resulted only in more side effects and dysphoria but little additional improvement in schizophrenic symptomatology. (After dose increase BPRS-Schizophrenia factor improved by only an additional 1.1 + 2.5 points (t = 1.58; p = n.s.), but BPRS-Depression worsened by -2.5 + 3.4 points (t = 2.71; p = 0.19) and BPRS-Withdrawal-Retardation worsened by -1.4 + 2.3 (t = 2.21; p = .047).

We cannot, of course, say that the 14 poor responders might not have worsened anyway during another four weeks of treatment at the same dose of haloperidol. To rule out the possibility that this worsening was not simply the continuation of trends prior to dose increase, we compared the change before the dose increase with the change after. A test for non-linerarity suggested that depression (t = 1.8; p = .09) and akinesia (t = 1.9; p = .008) worsened at an accelerated rate after dose increase.

These findings do not invalidate earlier work (35,36) in which we described florid exacerbations of psychosis secondary to EPS with high-potency neuroleptics. At that time, less was known about EPS and our patients experienced much higher EPS ratings. Similarly, the lack of a correlation between akinesia and depression ratings in this study does not, at least in our opinion, invalidate the notion of "akinetic depression" in our former work (37). The patients with "akinetic depression" (37) had higher akinesia ratings.

Out present findings show that patients at higher daily doses of haloperidol (20 mg orally) experienced more EPS (specifically akinesia and akathisia), and are inconsistent with the notion that higher doses of haloperidol cause less EPS of this type than low-doses (38,39). Uncontrolled observations can seem to support such a notion, for in any population clinical treatment will select out those patients (often chronic and relatively treatment-resistant) who tolerate high doses of potent neuroleptics remarkably well. If these patients are then contrasted with those who have a more usual extrapyramidal threshold, the illusion appears confirmed. We are not aware of any reports that show EPS at a low dose but not at high dose in the same patient.

The finding that a dose as low as 20 mg has apparent "psychotoxic" effects (higher drop-out rates, severe subjective distress due to akathisia, more akinesia and emotional withdrawal/blunted affect) may seem contrary to conventional thinking. Haloperidol, at least in the United States, is commonly prescribed aggressively, based on notion that it has a wide therapeutic index or margin of safety. In a busy in-patient service with a short length of stay and an understandable emphasis on gaining rapid antipsychotic effects, haloperidol has been claimed to be well tolerated and very effective at doses up to 100 mg daily (18-22). On

238

the other hand, if one considers EPS and dysphoric emotions, haloperidol has a narrow therapeutic index.

In conclusion, we have found that a daily oral dose of 20 mg of haloperidol is more effective for psychotic symptoms in the short-term treatment of psychotic, newly admitted male schizophrenic patients than 5 mg, and marginally superior to 10 mg. However, the 20 mg dose has more side effects, especially akinesia and akathisia, which seem to have psychopathological correlates of non-compliance, anxiety, dysphoria, withdrawal and apathy, leading to a complex balance of benefit versus risk. We suggest that, after the first week or so of treatment at daily doses of 10-20 mg of haloperidol or its equivalent, doses as low as 5 mg may be adequate. It is our impression that side effects of high doses of potent neuroleptics continue to be misinterpreted or overlooked and that the practical therapeutic index of such agents (dosage-ration of unwanted to therapeutic effects denoting margin of safety) is close to 1.0, that side effects (especially withdrawal and akinesia) are often sought as a desirable effect, and that these findings [together with other recent clarifications of neuroleptic dose-effect relations (15)] will lead to a more rational and safer use of these agents.

Plasma Level of HPL

In these same patients, a powerful curvilinear relationship was found between plasma HPL and clinical response when patients were given fixed doses of HPL (See Table 2; Figure 1).

The upper and lower limits of this proposed therapeutic window were 12 and 2 ng/mL, respectively. When plasma levels in relative nonresponders were pushed beyond 12 ng/mL (as in routine clinical practice), they, on balance, deteriorated. In particular, they became (relative to their status at the "therapeutic" plasma level) more dysphoric (BPRS-Depression factor = -2.6, SD 2.8, $p < .07$), more withdrawn (Withdrawal-Retardation factor = -1.3, SD 2.6, NS), and did not improve in global psychosis ratings (Schizophrenia + Paranoia factor = -0.83, SD 3.4, NS).

Similarly, when patients with plasma levels above 12 ng/mL had their plasma levels lowered, they, on balance, improved (BPRS-Schizophrenia = +1.25, $p < .05$; Withdrawal Retardation = +2.1, $p < .01$). No case deteriorated. Some improvement occurred in the sense that they no longer appeared overmedicated in terms of a personal sense of sedation and/or objective akinesia. In these cases a plasma level would only have confirmed what was already clinically apparent. Three cases with plasma levels above 12 ng/mL had become globally worse, developed a frantic agitation, and insisted on leaving the hospital before plasma levels could be lowered (2 of these cases had become profoundly depressed). Finally, 2 cases had developed delusions of bodily destruction which disappeared as the plasma level was lowered. We, along with Quitkin and colleagues (12), believe that delusions of bodily destruction can be a psychotic rationalization of neuroleptic toxicity.

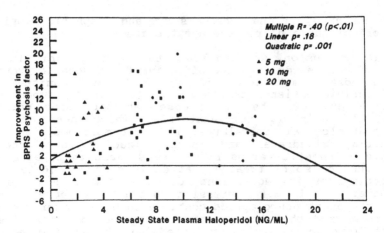

Figure 1.   Relationship Between Plasma HPL and Psychotic
            Symptoms

Table 2.   Regression Analysis Evaluating Linear and
            Curvilinear Relationships Between Plasma
            Haloperidol and Change in Brief Psychiatric Rating
            Scale (BPRS) Measures

	p. value		
	Linear	Quadratic	Cubic
Schizophrenia	.0481	.0069	
Depression	NS	.0455	
Paranoia	NS	.0093	
Withdrawal-Retardation	NS	NS	.0496
Total	NS	.0046	
Schizophrenia + Paranoia	NS	.0014	

Our proposed range of 2 to 12 ng/mL may not apply to the
chronic, treatment-refractory patient.   Certainly many such
patients are stabilized at plasma levels of HPL much greater
than 12 ng/mL.   The fact that such patients, to varying
degrees, can tolerate such high levels does not, however,
mean that this is their optimal plasma level.   Further, the
higher plasma levels in some treatment-refractory patients
are really utilized for "chemical restraint."   "Chemical
restraint," useful and necessary as it may be, was not a
desired endpoint in this sample; the patients in this sample
were reasonably cooperative and calm thought-disordered
schizophrenic men.   In such a sample our lower limit of 2.0
ng/mL is identical to the plasma HPL level (4.9 ± 2.9 ng/mL)
drawn at "neuroleptic threshold" by McEvoy and colleagues
(40).

CONCLUSION

Heretofore, dose-response analyses have focused
virtually exclusively on antipsychotic effects.   Dysphoric

responses to antipsychotic drugs have received little systematic inquiry. Yet they are important: no matter how much a patient improves behaviorally, in the end we have to pay attention to how he feels.

Dysphoric responses to medication erode the quality of life. Lack of vigor, ever present fatigue, lack of interest or motivation, muscular cramps and achiness, inner restlessness or impatience, an inability to feel comfortable in any body position, and vague complaints of "just not feeling right" are feeling states that schizophrenics often (correctly) attribute to their medications. Many of these dysphoric feeling states are extrapyramidally based and, since extrapyramidal symptoms in any one individual are always dose-dependent, a reduction in dosage often reduces a neuroleptic-induced dysphoria.

REFERENCES

1. Prien RF, Cole JO: High dose chlorpromazine therapy in chronic schizophrenia. Arch Gen Psychiatry 18: 481-495, 1968.
2. Prien RF, Levine J, Cole JO: High dose trifluoperazine therapy in chronic schizophrenia. Am J Psychiatry 126: 305-313, 1969.
3. Clark ML, Ramsey HR, Ragland RE, et al: Chlorpromazine in chronic schizophrenia: Behavioral dose-response relationships. Psychopharmacology (Berlin) 18: 260-270, 1970.
4. Clark ML, Ramsey HR, Rahhal DK, et al: Chlorpromazine in chronic schizophrenia. Arch Gen Psychiatry 27: 479-483, 1972.
5. Brotman RK, Muzekari LH, Shanken PM: Butaperizine in chronic schizophrenic patients: A double-blind study. Curr Ther Res 11: 5-8, 1969.
6. Carscallen HB, Rochman H, Lovegrove RD: High dosage trifluoperazine in schizophrenia. Can Psych Assoc J 13: 459-461, 1968.
7. Itil TM, Saletu B, Hu W, et al: Clinical and quantitative EEG changes at different dosage levels of fluphenazine treatment. Acta Psychiatr Scand 47: 440-451, 1971.
8. Itil TM, Keskiner A, Heinemann L, et al: Treatment of resistant schizophrenics with extreme high dosage fluphenazine hydrochloride. Psychosomatics 11: 4a56-463, 1970.
9. Simpson GM, Amin M, Kurk-Bartholini E, et al: Problems in the evaluation of the optimal dose of phenothiazine (butaperazine). Dis Nerv Syst 29: 478-484, 1968.
10. McClelland HA, Farquharson RG, Leyburn P, et al: Very high dose fluphenazine decanoate. A controlled trial in chronic schizophrenia. Arch Gen Psychiatry 33: 1435-1439, 1976.
11. Glodstein MJ, Rodnick EH, Evans JR, et al: Drug and family therapy in the aftercare treatment of acute schizophrenia. Arch Gen Psychiatry 35: 1169-1177, 1978.
12. Quitkin F, Rifkin A, Klein DF: Very high dosage vs standards dosage fluphenazine in schizophrenia. Arch Gen Psychiatry 32: 1276-1281, 1975.

13. Wijsenbeck H, Steiner M, Goldberg SC: Trifluoperazine: A comparison between regular and high doses. Psychopharmacology (Berlin) 36: 147-150, 1974.
14. Cole JO: Antipsychotic drugs: is more better? McLean Hosp J 7: 6-7, 1982.
15. Baldessarini RJ, Cohen BM, Teicher MH: Significance of neuroleptic dose and plasma level in the pharmacological treatment of psychosis. Arch Gen Psychiatry 45: 79-91, 1988.
16. Baldessarini RJ, Katz B, Cotton P: Dissimilar dosing with high-potency and low-potency neuroleptics. Am J Psychiatry 141: 748-752, 1984.
17. Gelenberg AJ, Bellinghausen B, Wojcik JD, Falk WE, Sachs GS: A prospective survey of neuroleptic malignant syndrome in a short-term psychiatric hospital. Am J Psychiatry 145: 517-518, 1988.
18. Neborsky R, Janowsky D, Nurison E, Munson E, Depry D: Rapid treatment of acute psychotic symptoms with high and low-dose heloperidol. Arch Gen Psychiatry 38: 195-201, 1981.
19. Eriksen SE, Hurt SW, Chang S: Haloperidol dose, plasma levels and clinical response: a double-blind study. Psychopharmacol Bull 14: 15-16, 1978.
20. Reschke R: Parental haloperidol for rapid control of severe, disruptive symptoms of acute schizophrenia. Dis Nerv Sys 35: 112-115, 1974.
21. Modestin J, Toffler G, Pia E, Greub E: Haloperidol in acute schizophrenia inpatients: a double-blind comparison of two dosage regimens. Pharmacopsychiatria 16: 121-126, 1983.
22. Donlon PT, Hopkin JT, Tupin JP, Wicks JJ, Wahaba M, Meadow A: Haloperidol for acute schizophrenia patients. An evaluation of three oral regimens. Arch Gen Psychiatry 37: 691-695, 1980.
23. Overall J, Gorham D: Brief Psychiatric Rating Scale. Psych Rep 10: 799-812, 1962.
24. Marder SR, Van Putten T, Mintz J, McKenzie J, Lebell M, Faltico G, May PRA: Costs and benefits of two doses of fluphenazine. Arch Gen Psychiatry 41: 1025-1029, 1984.
25. Teicher MH, Baldessazrini RJ: Selection of neuroleptic dosage. Arch Gen Psychiatry 42: 636-637, 1985.
26. Rifkin A, Quitkin F, Klein DF: Akinesia. Arch Gen Psychiatry 32: 672-674, 1975.
27. Rifkin A, Kane JM: Low-dose neuroleptic maintenance treatment of schizophrenia, in Kane JM (ed): Drug Maintenance Strategies in Schizophrenia. Washington, DC, American Psychiatric Press, Inc, pp 14-29, 1984.
28. Kane JM, Rifkin A, Woerner M, Reardon G, Kreisman D, Blumenthal R, Borenstein M: High-dose versus low-dose strategies in the treatment of schizophrenia. Psychopharmacology Bull 21: 533-537, 1985.
29. Drake RE, Erhrlich J: Suicide attempts associated with akathisia. Am J Psychiatry 142: 499-501, 1985.
30. Shear K, Frances A, Weiden P: Suicide associated with akathisia and depot fluphenazine treatment. J Clin Psychopharmacol 3: 235-236, 1983.
31. Weiden P: Akathisia from prochlorperazine. JAMA 253:635, 1985.
32. Van Putten T: The many faces of akathisia. Compr Psychiatry 16: 43-47, 1975.

33. Keckia WA: Violence as a manifestation of akathisia. JAMA 240:2185, 1978.
34. Shaw ED, Mann JJ, Weiden PJ, Sinsheimer LM, Brunn RD: A case of suicidal and homicidal ideation and akathisia in a double-blind neuroleptic crossover study. J Clin Psychopharmacol 6: 196-197, 1986.
35. Van Putten T, Mutalipassi LR, Malkin MD: Phenothiazine induced decompensation. Arch Gen Psychiatry 30: 102-105, 1974.
36. Van Putten T, Mutalipassi LR: Fluphenazine enanthale induced decompensation. Psychosomatics 16: 37-40, 1975.
37. Van Putten T, May PRA: "Akinetic depression" in schizophrenia. Arch Gen Psychiatry 35: 1101-1107, 1978.
38. Donlon PT: High dosage neuroleptic therapy: A review. Int Pharmacopsychiatry 11: 235-245, 1976.
39. Ayd FJ Jr: Haloperidol update: 1975. Proc R Soc Med 69: 14-22, (Suppl 1), 1976.
40. McEvoy JP, Stiller RL, Farr R: Plasma haloperidol levels drawn at neuroleptic threshold doses: A pilot study. J Clin Psychopharmacol 6: 133-137, 1986.

DRUG AND SUBJECT INFLUENCES ON MEASURES OF DEPRESSION AND

NEGATIVE SYMPTOMS IN SCHIZOPHRENIC PATIENTS

Richard Williams and J. Thomas Dalby

Departments of Psychiatry and Psychology
Calgary General Hospital and
University of Calgary, Calgary Alberta

INTRODUCTION

Kraepelin (1) and Bleuler (2) in their seminal writings pointed out that "genuine depression" frequently occurred in schizophrenic patients and suggested that dysphoric mood arose due to awareness of the disorder and its limiting influence on their life. In the formal development of mental disorder taxonomies a preference was expressed for orthogonal dimensions which implied that pure schizophrenia excluded depression. The sheer weight of clinical observations that depressive features occurred in schizophrenic patients was tentatively acknowledged in the designation of "Schizoaffective Disorder" and DSM-III-R (3) (p. 208) notes that this "represents one of the most confusing and controversial concepts in psychiatric nosology". Some investigators interpreted "depression-like" symptoms to be manifestations of Bleuler's symptoms, now rediscovered and referred to as "negative symptoms". These include affective flattening and alogia. Still others have suggested that depressed states are pharmacogenically induced (4-6). This latter idea has not received support in more recent and properly controlled investigations (7,8).

In the last decade there has been a renewed interest in depression in schizophrenic and a more general challenge to the view of independence in the classification and etiology of mental disorders (9). The clinical importance of addressing mood disturbance in schizophrenia is heightened by the reports of its prevalence (nearly 60% of one sample of schizophrenic patients had suffered a depressive syndrome during the course of their illness) (10), its relationship to poor outcome and risk of relapse (11,12), and its association with suicide (13,14).

In studying depression in schizophrenic individuals the first hurdle is identification since some symptoms which are associated with depression syndromes alone (e.g. flat affect) may not discriminate depression in schizophrenics.

Secondarily, one searches for variables which may influence the expression of mood disturbance.

Our study represents observations of naturalistic relationships that were uncovered in a group of schizophrenic patients. It is not an experimental study with random assignment to drug conditions and other controls. Nonetheless it probably represents an ecologically similar situation to that of many outpatient clinical settings.

METHOD

Subjects

Sixty patients meeting DSM-III criteria for chronic schizophrenia completed clinical testing for symptoms of depression and negative symptoms. This group, comprised of 37 men and 23 women, had a mean duration of 10.5 years (SD = 7.30) for their disorder with mean age of onset being 22.9 years (SD = 7.80). Their current mean age was 33.7 (SD = 10.13) years and they had completed a mean of 11.03 years (SD = 3.5) of education. The women were significantly older than the men (37.8 years versus 31.2 years; t (58) = 2.78, $\underline{P}$ <.05) but there was no difference in the duration of the illness. This phenomenon is accounted for by the fact that women in our sample, as is generally found (15), developed the first signs of their disorder an average of just over four years later than the men.

All subjects were attending a psychiatric day hospital treatment programme at the Calgary General Hospital.

Measures

We chose the two most common self-report measures of depression: the MMPI Depression Scale and the Beck Depression Inventory. The MMPI - D is composed of 60 true - false items and is heterogeneous in terms of item composition. Only eleven of the items in this scale were originally chosen on the basis of discriminating depressed patients from other psychiatric disorders. Other dimensions tapped by this scale have been referred to as psychomotor retardation, complaints about physical malfunctioning, mental dullness and brooding which also appear in other mental disorders. On the other hand, the Beck Depression Inventory (BDI) shows substantially higher internal consistency of items. This test consists of 21 items and each item consists of four or five statements ranked in order of severity of the specific depression manifestation. It is generally considered the best self-report measure of depression severity. We also administered the Beck Hopelessness Scale (BHS)(16) as some (13) have suggested that it is this element which is the critical factor in determining suicide potential in schizophrenics. The BHS is comprised of 20 true/false questions with high internal consistency which focus on hope for the future (e.g. "All I can see ahead of me is unpleasantness rather than pleasantness").

Finally we assessed all patients using the Scale for the Assessment of Negative Symptoms (SANS)(17) as some question had been raised regarding the relationship between depression and negative symptoms in schizophrenia. In our analysis we employed summary scores which summed global ratings of five symptom complexes (affective flattening or blunting, alogia, avolition-apathy, anhedonia-asociality and attentional impairment).

RESULTS

Table 1 presents a correlation matrix between the four measures for the entire sample.

Table 1. Correlation Matrix of Scales

	BDI	BHS	SANS
MMPI-D	.30*	.42**	-.09
BDI		.58**	.15
BHS			.15
SANS			

* p<.05
**p<.01

Relationship Between Scales

Although the two depression scales were significantly and positively correlated, the magnitude of the relationship was less than has previously been reported with depressed patients without schizophrenia (18,19). Both depression scales were also significantly related to the dimension of hopelessness as measured by the BHS. Striking is the lack of relationship between depression or hopelessness and negative symptoms. Andreasen (17) suggests that negative symptoms, such as poverty of speech or physical anergia, will be seen in patients with other diseases particularly depression but also notes that approaches to diagnoses should be polythetic. There is also a source difference between the scales, with depression and hopelessness reflecting the patients own perceptions while the SANS relies upon a rating by a psychiatrist.

Levels of Depression

On the Beck Depression Inventory a significant sex difference emerged with females scoring higher (mean score = 17.48) than males (mean score = 10.81), t (58) = 2.38, p < .05. Table 2 represents frequency data reflecting the distribution of scores on the BDI.

On the MMPI-D no significant sex difference was seen (mean raw score: Males = 26.7; Females = 29.17). Using more typical T-scores, however, more men than women scored over 70 T (which is two standard deviations above the mean). (Table 3) Men, as a group, tend to acknowledge fewer depressive

247

Table 2. Distribution of BDI Scores

	Males	Females	Total
0 - 9 (no depression)	19	7	26
10 - 15 (dysphoria)	8	4	12
16+ (depression)	10	12	22

symptoms and therefore it requires fewer test items to be endorsed to enter the clinically significant range ($\geq$ 70T). Thus it is important to understand the psychometric properties of the scale used in assessing depression in this population.

Table 3. Distribution of MMPI-D Scores

	Males	Females	Total
<70T	12	13	25
>70T	25	10	35

Given the heterogeneity of the items in the MMPI-D we compared the ranking of the ten most frequently cited items in men and women. Eight of the ten most frequently endorsed items of women were those which factor analyses have identified as "subjective depression" (e.g. "I wish I could be as happy as others seem to be" - True) while males only endorsed two of these items in the top ten. Males tended to focus more on physical complaints (e.g. "I seldom worry about my health" - False) and aggression (e.g. "At times I feel like smashing things" - True).

On measures of hopelessness (Beck Hopelessness Scale) and negative symptoms (SANS), no sex differences were observed. In multiple regression analyses and correlations no relationships were uncovered between age, length of disorder, education, age of onset and scores on the BHS or the SANS.

Drug influences

This group of patients had been stabilized on a variety of neuroleptic medications, most receiving depot forms. Linear analyses of drug dosage required the drugs to be analyzed separately. The largest number of patients (N = 31) were receiving Flupenthixol Decanoate. The mean dosage for this group was 77.03 mg/2 weeks (median = 50 mg). Dose was not significantly correlated with the depression measures or the hopelessness scale. However, it was positively and significantly related to the SANS (r = .285, p <.05) Thus higher doses were related to more observed negative symptoms. We then examined dose by range (low, normal, high). These were defined as <20 mg/2 weeks (low), 20 - 80 mg/2 weeks (normal) and >80 mg/2 weeks (high). As seen in Table 4 only the high dose group had an elevated SANS rating.

Table 4. Distribution of SANS Scores by Dose

	N	Mean	S.D.
Low	5	7.20	6.42
Normal	17	8.71	4.95
High	9	12.44	3.78

Females and males in this group did not differ in dose. Separate correlations by sex did not uncover specific relationships related to this independent variable interacting with drug dose.

Fourteen subjects were receiving an antiparkinsonian agent (Cogentin) and in comparisons with those patients not receiving this medication no differences were noted on the dependent measures.

The second largest group was comprised of patients receiving Fluphenazine Decanoate (N = 15). They had been receiving a mean dose of 27.75 mg/2 weeks (median = 25 mg/2 weeks). No significant relationships emerged between drug dose and the four dependent measures. The lack of a dose related effect on SANS can be explained by the relatively moderate dosage level with only two patients receiving greater than 50 mg/2 weeks. The variance of dosage was also rather restricted with this drug.

In comparing these two depot neuroleptics, no differences were noted between them on dependent variables when sex was accounted for.

The remaining patients received a variety of medications (Haldol LA, Stelazine, Chlorpromazine) and when this "other drug" group was compared to the Flupenthioxol and Fluphenazine groups again, no differences were seen on dependent measures.

DISCUSSION

Measurement of depression

As the appropriate clinical measure of depression in this population had not been comprehensively explored we chose to administer multiple measures of depressive-related symptoms. We felt that it was important to explore the subjective thoughts of the patients rather than rely only on behavior ratings as these may be deceptive to the observer (10).

With regard to the use of the MMPI with schizophrenics there is little doubt that this has been the most frequently used psychometric instruments with this population. Many studies have reported that the two highest scales in the schizophrenic population are Sc (Schizophrenia) and D

(Depression) (20,21) suggesting a high incidence of depression in schizophrenia. We have previously reported (22) that these scales are relatively independent with no significant correlation between them. A major review of the MMPI with schizophrenics (23) points out the many advantages of this instrument but cautions that it should be viewed as providing the clinician with probabilistic statements which can then be explored further by more specific methods. Because of its widespread use the MMPI will continue to be a source of information on depression in schizophrenics. Our findings emphasize the importance of understanding the psychometric properties of this scale and the heterogeneity of its internal structure. Future research may explore the subscales embedded in the Depression scale of the MMPI to delineate the dimensions of depression prominent in schizophrenic patients at different times during the course of their schizophrenic disorder.

The Beck Depression Inventory indicated that using traditional cut-off scores developed for depressed patients that 37% of our sample would fall into the depressed category. It also emphasized a strong sex-difference in the identification of more unitary depression symptoms than the MMPI-D. Studies of other populations with the BDI have revealed inconsistent sex differences, but when such differences are noted females typically score higher on this instrument (24). The level of depression in our schizophrenic sample as measured by this instrument, appears much lower (24) and slightly higher (25) than others. Differences in the course of the disorder of subjects may account for differences (26). We favor the BDI for the simplicity of administration, generally good psychometric properties (particularly discriminant validity) and its demonstrated sensitivity to clinical change (27).

The hopelessness inventory correlated positively with other depression measures and its independent clinical utility is uncertain. Because this emotional element may be central to suicidal behavior in schizophrenics (13) we would suggest the employment of this scale in large scale predictive studies of suicidal behavior. The inventory's weakness at this point lies in the lack of well established criteria for the diagnosis of "hopelessness". We found a bimodal distribution with the first mode on a score of two and the second on seven. There were very few very high scores (maximum on this scale is 20) even though there is a very high internal consistency of items on this measure.

The SANS was unrelated to measures of depression or hopelessness thereby rejecting the notion that "depressive-like" symptoms in schizophrenia are really only negative symptoms. There may be some relationship between motor elements of depression and negative symptoms but this would be a superficial and unimportant concern.

Drug influences

We found no significant relationship between drug dosage and symptoms of depression, confirming most recent analyses of this relationship (28). Early findings of a neuroleptic induced "depression" may have been actually reporting symptoms which now would be classified as "negative symptoms". As Baldessarini, Cohen and Teicher (29) have stated, the clinical practice with neuroleptics, until most recently, was to encourage high-potency agents in high doses to increase the degree and speed of therapeutic response. We have pointed out that these high doses are related to increases in negative symptoms.

CONCLUSIONS

We reiterate Kraepelin and Bleuler's concern with "genuine depression" in schizophrenia and reject the notion that depression is caused by neuroleptics or that depression is the manifestation of only "depressive-like" negative symptoms in this population. We examined the differential usefulness of types of measures of depression with schizophrenic patients. We also point out the substantial sex differences in the level of depression and the qualitative expression of depression between the sexes. Female schizophrenics express more typical "depressed mood" features while males (as has been noted elsewhere) (24) become more irritable and short tempered and focus on somatic complaints.

The treatment of depression in schizophrenic patients is a neglected aspect of care. The paucity of research on the effectiveness of antidepressive pharmacological agents with this patient population reinforces this opinion. Because of this therapeutic oversight it may be that the high incidence of substance abuse in schizophrenic patients (30) reflect attempts to self-treat depressive mood.

Finally we postulate that, like other depressed groups, depression in schizophrenics arises from multiple sources. Our previous finding in this area relating intelligence and depression (22) would suggest that cognitive variables are important mediators of depressed mood and therefore amenable to alteration through psychotherapy.

ACKNOWLEDGEMENT

This study was supported in part by grants from the Ian Douglas Bebensee Foundation and Merrell Dow Pharmaceuticals (Canada) Inc.

REFERENCES

1. Kraepelin E: Dementia Praecox and Paraphrenia. Edinburgh, Livingstone, 1919.
2. Bleuler E: Dementia Praecox or the Group of Schizophrenias. New York, International Universities Press, 1950.

3. American Psychiatric Association: <u>Diagnostic and Statistical Manual, Third Edition, Revised</u>. Washington, D.C., APA, 1987.
4. De Alarcon R, Carney MWP: Severe depressive mood change following slow-release intramuscular fluphenazine injection. <u>Br Med J</u> 3: 564-567, 1969.
5. Floru L, Heinrich K, Wittek F: The problem of post-psychotic schizophrenic depressions and their pharmacological induction. <u>Int Pharmacopsychiatry</u> 10: 230-239, 1975.
6. Keskiner A: A long-term follow-up of fluphenazine enanthate treatment. <u>Curr Ther Res</u> 15: 305-313, 1973.
7. Hogarty GE, Munetz MR: Pharmacogenic depression among outpatient schizophrenic patients: A failure to substantiate. <u>J Clin Psychopharmacol</u> 4: 17-24, 1984.
8. Wistedt B, Palmstierna T: Depressive symptoms in chronic schizophrenic patients after withdrawal of long-acting neuroleptics. <u>J Clin Psychiatry</u> 44: 369-371, 1983.
9. Crow TJ: The continuum of psychosis and its implication for the structure of the gene. <u>Br J Psychiatry</u> 149: 419-429, 1986.
10. Martin RL, Cloninger CR, Guze SB, Clayton PJ: Frequency and differential diagnosis of depressive syndromes in schizophrenia. <u>J Clin Psychiatry</u> 46 (11, Sect 2): 9-13, 1985.
11. Glazew W, Prusoff B, John K, et al: Depression and social adjustment among chronic schizophrenia outpatients. <u>J Nerv Ment Dis</u> 169: 712-717, 1981.
12. Herz MI, Melville C: Relapse in schizophrenia. <u>Am J Psychiatry</u> 137: 801-805, 1980.
13. Drake RE, Gates C, Cotton PG: Suicide among schizophrenics: a comparison of attempters and completed suicides. <u>Br J Psychiatry</u> 149: 784-787, 1986.
14. Black DW, Winokur G, Warrack G: Suicide in schizophrenia: The Iowa Record Linkage Study. <u>J Clin Psychiatry</u> 46 (11, Sect 2): 14-17, 1985.
15. Loranger AW: Sex difference in age at onset of schizophrenia. <u>Arch Gen Psychiatry</u> 46: 157-161, 1984.
16. Beck AT, Weissman A, Lester D, Trexler L: The measure of pessimism: The Hopelessness Scale. <u>J Consult Clin Psychol</u> 42: 861-865, 1974.
17. Andreasen NC: Negative symptoms in schizophrenia: Definition and reliability. <u>Arch Gen Psychiatry</u> 39: 784-788, 1982.
18. Seitz R: Five psychological measures of neurotic depression: A correlation study. <u>J Clin Psychol</u> 26: 504-505, 1970.
19. Turner JA, Romano JM: Self-report screening measures for depression in chronic pain patients. <u>J Clin Psychol</u> 40: 909-913, 1984.
20. Holland TR, Levi M, Watson CG: MMPI basic scales vs. two-point codes in the discrimination of psychopathological groups. <u>J Clin Psychol</u> 37: 394-396, 1981.
21. Rosen A: Differentiation of diagnostic groups by individual MMPI scales. <u>J Consult Psychol</u> 22: 453-457, 1958.

22. Dalby JT, Williams R: Depression in schizophrenia: The influence of intelligence. <u>Can J Psychiatry</u> 32: 816-817, 1987.
23. Walters GD: The MMPI and schizophrenia: A review. <u>Schizoph Bull</u> 9: 226-246, 1983.
24. Kleinke CL: Comparing depression-coping strategies of schizophrenic men and depressed and non-depressed college students. <u>J Clin Psychol</u> 40: 420-426, 1984.
25. Johnson DAW: Studies of depressive symptoms in schizophrenia I. The prevalence of depression and its possible causes. <u>Br J Psychiatry</u> 139: 89-101, 1981.
26. Leff J, Tress K, Edwards B: The clinical course of depressive symptoms in schizophrenia. <u>Schizophrenia Research</u> 1: 25-30, 1988.
27. Rush AJ, Khatami M, Beck AT: Cognitive and behavior therapy in chronic depression. <u>Behavior Therapy</u> 6: 398- 404, 1975.
28. Siris SG, Strahan A, Mendeli J, Cooper TB, Casey E: Fluphenazine decanoate dose and severity of depression in patients with post-psychotic depression. <u>Schizophrenia Research</u> 1: 31-35, 1988.
29. Baldessarini RJ, Cohen BM, Teicher MH: Significance of neuroleptic dose and plasma level in the pharmacological treatment of psychoses. <u>Arch Gen Psychiatry</u> 45: 79-91, 1988.
30. Schneier FR, Siris SG: A review of psychoactive substance use and abuse in schizophrenia: Patterns of drug choice. <u>J Nerv Ment Dis</u> 175: 641-652, 1987.

Back row (standing) left to right: Pierre Flor-Henry, David Romney, Donald Addington, Gerald McDougall, Richard Williams, Barry Jones, Theodore Van Putten, Nady el-Guebaly, Jerry Westermeyer, David Owens, Peter McGuffin, Michael Pogue-Geile. Front row (seated) left to right: Jean Addington, Martin Harrow, Paul Green, Thomas Dalby, Samuel Siris, Stephen Bartels, Alec Roy, Ming Tsuang, Donald Johnson, Keith Nuechterlein, Jacqueline Samson

CONTRIBUTORS

Donald Addington, M.B., B.S., M.R.C. Psych., F.R.C.P.(C.)

Department of Psychiatry
Foothills Hospital
1403 - 29th Street NW
Calgary, Alberta, Canada
T2N 2T9
and
Department of Psychiatry
University of Calgary

Jean Addington, Ph.D.

Holy Cross Hospital
2210 - 2nd Street SW
Calgary, Alberta, Canada
T2S 1S6

Stephen J. Bartels, M.D.

West Central Services, Inc.
2 Reservoir Road
Hanover, NH 03755
USA
and
Department of Psychiatry
Dartmouth Medical School

Ian F. Brockington, M.D., M.R.C.Psych.

Department of Psychiatry
University of Birmingham
Queen Elizabeth Hospital
Birmingham, England
B15 2TH

Nicole Chabert, M.A.

Veterans Administration Medical Center
11301 Wilshire Blvd., Building 210C
Los Angeles, California 90073
USA

J. Thomas Dalby, Ph.D.

Department of Psychology
Calgary General Hospital
841 Centre Avenue East
Calgary, Alberta, Canada
T2E 0A1
and
Departments of Psychiatry and Psychology
University of Calgary

Robert E. Drake, M.D., Ph.D.

West Central Services, Inc.
2 Reservoir Road
Hanover, NH 03755
USA
and Department of Psychiatry
Dartmouth Medical School

Anne E. Farmer, M.D., M.R.C.Psych.

Department of Psychological Medicine
University of Wales College of Medicine
Heath 46. Park, Cardiff
Wales
CF4 4XN

Pierre Flor-Henry, M.B., Ch.B., M.D., D.P.M., F.R.C.Psych.,
C.S.P.Q.

Alberta Hospital Edmonton
Box 307
Edmonton, Alberta
T5J 2J7
and
Department of Psychiatry
University of Alberta

Michael F. Green, Ph.D.

UCLA Research Center
Box A
Camarillo, CA 93011
USA

Martin Harrow, Ph.D.

Department of Psychology
Michael Reese Hospital and Medical Center
Lake Shore Drive at 31st Street
Chicago, IL 60616
USA
and
Departments of Psychiatry and
Behavioral Sciences
University of Chicago

256

Ian Harvey, M.B., M.R.C.Psych.
Kings College
University of Longon
London, England

Donald A.W. Johnson, M.D., M.Sc., F.R.C.Psych.

University Hospital of South Manchester
West Didsbury
Manchester, England
M20 8LR

Eve C. Johnstone, M.D., F.R.C.P., F.R.C.Psych., D.P.M.

Northwick Park Hospital and Clinical Research Centre
Watford Road, Harrow
Middlesex, HA1 3VJ
England

Barry Jones, M.D., F.R.C.P.(C.)

Department of Psychiatry
University of Ottawa
and
Royal Ottawa Hospital
1145 Carling
Ottawa, Ontario
K1Z 7K4

Stephen R. Marder, M.D.

Veterans Administration Medical Center
11301 Wilshire Blvd., Building 210C
Los Angeles, CA 90073
USA

Gerald McDougall, M.D., F.R.C.P.(C.)

Department of Psychiatry
Foothills Hospital
1403 - 29th Street NW
Calgary, Alberta
T2N 2T9
and
Department of Psychiatry
University of Calgary

Peter McGuffin, M.B., Ph.D., F.R.C.P., M.R.C.Psych.

Department of Psychological Medicine
University of Wales College of Medicine
Heath Park, Cardiff
Wales
CFN 4XN

Jim Mintz, Ph.D.

Department of Psychiatry and Bio-Behavioral Sciences
UCLA
Los Angeles, CA
USA

Keith H. Nuechterlein, Ph.D.

Department of Psychiatry and Biobehavioral Sciences
UCLA
Los Angeles, CA   90024
USA

David G.C. Owens, M.D.(Hons.), F.R.C.P., M.R.C.Psych.

Northwick Park Hospital and Clinical Research Centre
Watford Road, Harrow
Middlesex, HA1 3VJ
England

Michael F. Pogue-Geile, Ph.D.

Departments of Psychology and Psychiatry
University of Pittsburgh
4015 O'Hara Street
Pittsburgh PA 15260
USA

David Romney, Ph.D.

Department of Educational Psychology
University of Calgary
2500 University Drive
Calgary, Alberta, Canada
T2N 4N1

Alex Roy, MB

National Institutes of Mental Health,
Bldg. 10, R3B-19
9000 Rockville Pike
Bethesda, MD   20892
USA

Jacqueline A. Samson, Ph.D.

Brockton/West Roxbury Veterans Administration
Medical Center
940 Belmont Street
Brockton, MA   02401
USA

Samuel G. Siris, M.D.

Hillside Hospital
P.O. Box 38
Glen Oaks, NY 11004
USA

Bryan Tanney, M.D., F.R.C.P.(C.)

Department of Psychiatry
Calgary General Hospital
841 Centre Avenue East
Calgary, Alberta
T2E 0A1
and
Department of Psychiatry
University of Calgary

William Torrey, M.D.

Department of Psychiatry
Dartmouth Medical School
Hanover, NH
USA

Ming Tsuang, M.D., Ph.D., D.Sc.

Department of Psychiatry
Harvard Medical School
and
Brockton/West Roxbury VA Medical Center
940 Belmont Street
Brockton, MA 02401
USA

Theodore Van Putten, M.D.

Veterans Administration Medical Center
11301 Wilshire Boulevard
Building 210C
Los Angeles CA 90073
USA

Joseph Ventura, M.A.

Department of Psychiatry and Biobehavioral Sciences
UCLA
Los Angeles, CA 90024
USA

Di Vosburgh, B.N.

Department of Psychiatry
Foothills Hospital
1403 29th Street NW
Calgary, Alberta, Canada
T2N 2T9

Jerry Westermeyer, Ph.D.

Department of Psychology
Michael Reese Hospital and Medical Center
Lake Shore Drive at 31st Street
Chicago, IL 60616
USA
and
Department of Psychiatry
University of Chicago

Maureen Williams, M.A.

Kings College
University of London
London, England

Richard Williams, M.B., B.S., M.R.C.Psych., M.Phil.,
    F.R.C.P.(C.)

Psychiatric Day Hospital
Calgary General Hospital
841 Centre Avenue East
Calgary, Alberta, Canada
T2E 0A1
and
Department of Psychiatry
University of Calgary

Alexander Young, B.S., M.D.
Brockton/West Roxbury Veterans Administration
Medical Center
940 Belmont Street
Brockton, MA 02401
USA

INDEX